Scanning Force Microscopy

REVISED EDITION

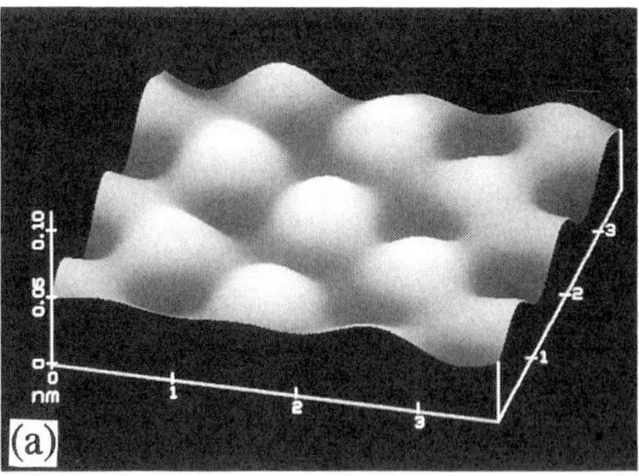

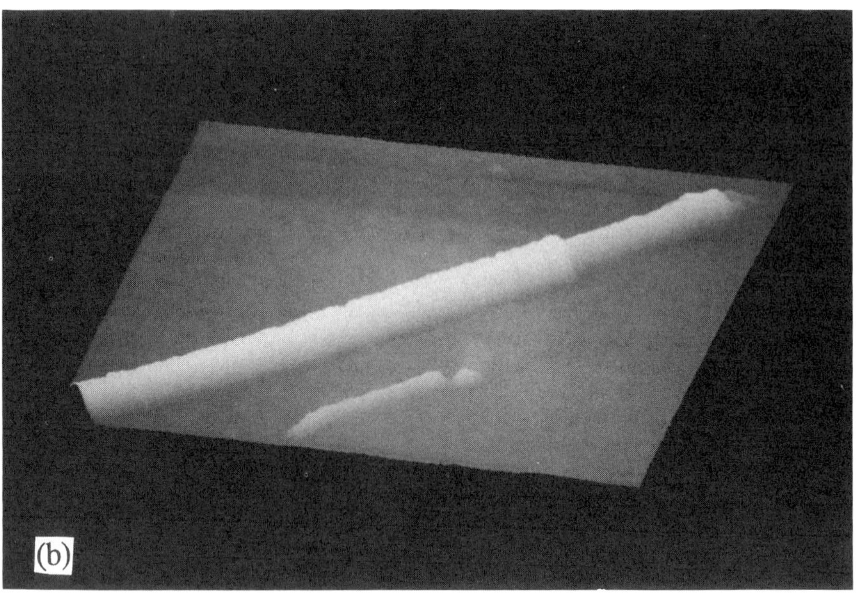

Atomic force microscopy image of (a) C_{60} molecules deposited on a gold substrate having a diameter of 1 nm and (b) carbon nanotubes having diameters of 14, 21, and 28 nm, obtained from the author's labs.

Scanning Force Microscopy

With Applications to Electric, Magnetic and Atomic Forces

REVISED EDITION

Dror Sarid
Optical Sciences Center
University of Arizona-Tucson

New York Oxford
OXFORD UNIVERSITY PRESS
1994

Oxford University Press

Oxford New York Toronto
Delhi Bombay Calcutta Madras Karachi
Kuala Lumpur Singapore Hong Kong Tokyo
Nairobi Dar es Salaam Cape Town
Melbourne Auckland Madrid

and associated companies in
Berlin Ibadan

Published by Oxford University Press, Inc.,
200 Madison Avenue, New York, New York 10016

Library of Congress Cataloging-in-Publication Data
Sarid, Dror.
Scanning force microscopy : with applications to electric,
magnetic, and atomic forces / Dror Sarid. — Rev. ed.
p. cm. (Oxford series in optical and imaging sciences ; 5)
Includes bibliographical references and index.
ISBN 0-19-509204-X
1. Scanning force microscopy.
2. Surfaces (Physics)
I. Title.
II. Series.
QH212.S32S27 1994 502'.8'2 — dc20 94-3081

9 8 7 6 5 4 3 2 1

Printed in the United States of America
on acid-free paper

To Lea, Uri, and Rami

Preface to the Revised Edition

High interest, accompanied by major advances in the technology of scanning force microscopy, which has taken place since this book first appeared early in 1991, warranted the preparation of a second print. Indeed, the tenth anniversary of Scanning Tunneling Microscopy, celebrated at the Sixth International Conference on Scanning Tunneling Microscopy in Interlaken, Switzerland August 12-16, 1991, with more than one thousand participants, produced three volumes of papers that attest to the ever-growing interest in this technology. As the field of scanning force microscopy has matured, a gradual shift in gears has taken place, from activities involving the development of instruments to their use as probes in a large rainbow of disciplines. Therefore, this new print, which includes all the material contained in the first print, additionally has the latest available list of new references that deal with electric, magnetic, and atomic force microscopy. It should also be noted that several scanning force microscopy related reviews have appeared, some of which present detailed information on specialized topics. For example, see Binnig (1992), den Boef (1991), Dürig et al. (1992), several articles appearing in *Scanning Tunneling Microscopy II*, edited by Wiesendanger and Güntherodt (1992), and in particular, a book by C. Julian Chen titled *Introduction to Scanning Tunneling Microscopy*, which treats the role of forces in scanning tunneling microscopy as well as atomic force microscopy.

Thanks are due to Sam Howells for careful reading of the book, and Linda Schadler and Meredith Binder for editorial revisions and preparation of the camera-ready copy.

Tucson, Arizona
November 1993

Preface

In the scanning tunneling microscope (STM), invented by Gerd Binnig and Heini Rohrer in 1982, an atomically sharp tip is placed sufficiently close to the surface of a conducting sample that tunneling of electrons between the two is possible. The tunneling current as a function of position of the tip across the sample provides an image that reflects the electronic structure of the uppermost atoms at the surface of the sample. The images, which can have atomic resolution, give a measure of the local density of states of the electrons whose energy is at the Fermi level. The atomically short depth of focus of the STM distinguishes it from scanning or transmission electron microscopes, whose high-resolution images derive from the projection of the chemical structure of a sample along the direction of their electron beam. Also, in contrast to electron microscopy, the STM, through the atoms at the apex of its probing tip, does exert a force on the surface of the sample which is of the same order of magnitude as that of interatomic forces.

This new effect gave rise to a novel direction put to good use by Gerd Binnig, Calvin Quate, and Christopher Gerber in 1986 when they published the first atomic-force-microscope (AFM) results. Here, instead of using an STM tip whose direction is normal to the surface of the sample, they positioned it in an almost parallel direction so that its sharp edge was just above the surface. The tip, acting as a cantilever, exerted a force on the sample the same way that the STM tip does, except that now the minute deflections of the lever with its force-sensing edge were of importance. To measure the deflection of the cantilever, they used an STM tip that could resolve cantilever deflections as small as 10^{-4} Å. The mechanical properties required of such a cantilever were quite simple. Since the mass and frequency of vibration of a typical atom are $\omega \simeq 10^{13}$ rad and $m = 10^{-25}$ kg, the atom equivalent spring constant is $k = \omega^2 m = 10$ N/m. The spring constant of an AFM cantilever should therefore be smaller than 10 N/m to avoid damaging the surface of the sample. Small metallic wires with a diameter on the order of 10 micrometers and a length of several hundred micrometers will satisfy that requirement. A much better choice, however, turned out to be microfabricated Si, SiO_2, and Si_2O_3 cantilevers having a spring constant on the order of 1 N/m and a resonance frequency of several tens of kHz, high enough to enable rapid raster scanning. By displaying the deflection of the cantilever as a function of position across a sample that is not necessarily conducting, they obtained atomically resolved images. In contrast to tunneling, where only electrons with momentum $k_+ = k_F$ contribute appreciably to the tunneling current, atomic forces involve electrons for $k < k_F$ (Dürig et al. 1986).

The next step in the development of the AFM came about when other methods for monitoring the deflection of its cantilever were invented. Although the AFM cantilever could be made sufficiently soft to avoid damaging the biomaterials (acting force $F \simeq 10^{-9}$ N), better methods were still needed because the tunneling tip that monitored its deflection exerted a large force on it ($F \simeq 10^{-7}$ N). Also, the existence of contamination layers between the AFM cantilever and the tunneling tip made this monitoring method unreliable. Optical methods proved to be a better alternative to the STM tip for monitoring the deflection of the AFM cantilever (McClelland et al. 1987, Martin et al. 1987), because the force they exert on the cantilever is negligible, the deflection that they measure averages the rough surface of the cantilever, and they are not sensitive to contamination layers on the cantilever. Indeed, most AFMs currently employ optical sensors of some sort.

The two complementing technologies, the STM and the AFM, fulfilled such a needed technology that their understanding and use has grown exponentially since their inception. Yet to assess their potential one should recognize the capability of competing technologies. Being concerned here with atomic force microscopy, we should look on those technologies that probe surface topography and electrostatic and magnetostatic stray fields across a surface. The technology used for characterizing surfaces, prior to the inception of the AFM, primarily consisted of stylus profilometers (Young et al. 1972, Teague et al. 1982) that have demonstrated lateral and vertical resolutions of 1000 Å and 10 Å, respectively, and the scanning capacitive microscope that demonstrated lateral and vertical resolutions of 5000 Å and 2 Å, respectively. Magnetostatic stray fields across surfaces can be measured by various methods. Among them, Bitter decoration outlines domain boundaries yet leaves unacceptable residue; scanning electron microscopy with polarization analysis (SEMPA) requires vacuum, clean surfaces, and lengthy accession times; and Lorentz microscopy requires vacuum and thinning of the sample for transmission of electrons. For probing electrostatic stray fields on a semiconductor structure, for example, one can use a scanning electron microscope that bombards it with high-energy electrons. Vacuum requirements and the charging of the semiconductor structure, however, make it a less attractive method.

Unlike the AFM that usually operates in the tip-sample contact mode, electrostatic or magnetostatic fields are probed in the noncontact mode. Vibrating the cantilever for these applications can increase the sensitivity by several orders of magnitude because (1) the operating frequency is removed from the region where $1/f$ noise has a significant contribution, (2) use of a phase-sensitive detection method increases the signal-to-noise ratio, and (3) it is possible to use lever resonance enhancement to increase the sensitivity. By operating on the steepest slope of the resonance of a lever, Martin et al. (1987) demonstrated a sensitivity to forces of 3×10^{-13} N and force derivatives of 1.5×10^{-4} N/m. Using a feedback-driven lever, Albrecht et al. (1990a) demonstrated a sensitivity to force derivatives of 9×10^{-5} N/m, which could be made as small as 10^{-8} N/m. Also, for systems based on interferometry, operating with a vibrating lever provides two degrees of freedom, one associated with the ac output of the photodetector and the other one associated with its dc output. In a magnetic force microscope, for example, the two degrees of freedom make it possible to partially isolate topographic and magnetostatic information (Schönenberger and Alvarado 1990a and Schönenberger et al. 1990). With the electric force microscope, the application of both dc and ac fields between tip and sample makes it possible to distinguish between the polarity of charges deposited on the surface of a sample (Stern et al. 1988).

Excellent reviews of topics in atomic force microscopy have already been published. For the contact mode of operation, Binnig, Quate, and Gerber's original paper (1986) can serve as the basic introduction to AFM, as well as the papers by Martin et al. (1987), McClelland et al. (1987), and Hansma et al. (1988). Other reviews have been given by McClelland (1987), Wickramasinghe (1989 and 1990), Rugar and Hansma (1990), Sarid and Elings (1991), Zasadzinski et al. (1988), Landman et al. (1990), and Schönenberger and Alvarado (1990a). Also, the book by Israelachvili is an invaluable source of information on intermolecular and surface forces. Now, the rapid growth of these two technologies, and in particular atomic force microscopy, put a stressful drive on part of each and every involved researcher, directing their effort at advancing these technologies at the fastest possible rate. This author felt that at this point in time it is worthwhile to pause for a moment, reflect on past accomplishments, and attempt to arrange the material appearing in the literature according to specific categories. The purpose of this book, therefore, is to try and present a unified view of the rapidly growing field of atomic force microscopy. The title of the book reflects a somewhat broader range than atomic force microscopy that has since branched into two directions primarily: contact and noncontact interactions. To the first direction belongs the original atomic force microscope that maps tip-sample interaction at atomically close distances. The second direction includes electric and magnetic force microscopes where electrostatic and magnetostatic interactions are expected to dominate over atomic force interactions. The noncontact type of microscope has its tip held farther from the surface of the sample, where, in contrast to the contact mode of operation, the tip can be made to vibrate and increase the sensitivity to forces. Because the book deals with three related microscopies, it was more meaningful to title it Scanning Force Microscopy (SFM), rather than Atomic Force Microscopy. The division into three separate microscopies is also in line with the complexity of the underlying physics. Tip-sample electrostatic interactions can be analyzed using first-order theory because one is dealing with electric monopoles. Next, in order of complexity, are magnetostatic interactions that deal with dipoles and cooperative phenomena such as ferromagnetism and magnetic domains. Atomic force interactions are the more complex ones and require molecular-dynamics simulations for their detailed understanding.

The presentation of the book is a mixture of a text book and a review article. An attempt was made to present a theoretical background of scanning force microscopy that can be followed from first principle, where detailed experimental information is incorporated into each chapter. The book is divided into three parts: (a) Levers and Noise, (b) Scanning Force Microscopes, and (c) Scanning Force Microscopy. The division of the book into these parts is based on the distinction between the three aspects of scanning force microscopy, namely, mechanics, instrumentation, and physics. The first part consists of three chapters that deal with mechanical properties of levers, resonance enhancement, and sources of noise. The second part consists of seven chapters that discuss seven implementations of scanning force microscopes. The third part consists of three chapters that deal with the physics and applications of scanning force microscopy in the order of their complexity: electrostatic, magnetostatic, and atomic-force interactions.

Chapter 1, Mechanical Properties of Levers, treats the classical problem of vibrating cantilevers, denoted in short, levers. Here we deal with strain and stress, moments, spring constants and lumped systems, Rayleigh and classical solutions to vibrating levers and their normal modes, and conclude with examples that give expressions useful for the design of levers. Chapter 2, Resonance Enhancement, describes the properties of bimorphs used for positioning and vibrating a lever and introduces the concept of an effective spring constant. Following is a treatment of three basic SFM configurations used in noncontact modes where resonance enhancement of levers is of advantage. The configurations consist of a (1) bimorph-driven lever, (2) sample-driven lever, and (3) tip-driven lever. A voltage-driven lever is treated separately in the section dealing with electric force microscopy where it belongs naturally. Chapter 3, Sources of Noise, presents a general discussion of noise consisting of shot noise, resistor thermal noise, laser intensity and phase noise, thermally induced lever noise, bimorph noise, and lever noise-limited signal-to-noise ratio. Each of the next seven chapters describes different implementations of a scanning force microscope, presenting their theory, noise considerations, and performance. The seven implementations employ the following lever-deflection detection methods: (1) tunneling, (2) capacitance, (3) optical homodyne, (4) optical heterodyne, (5) laser-diode feedback, (6) optical polarization, and (7) optical deflection. The first two use electronics as a means of detection while the five others use optical methods. The last three chapters deal with the theory, principles of operation, and performance of scanning force microscopy. Chapter 11 addresses electric field microscopy, which images surface topography and potentials across a conducting surface. The electrostatic fields consist of stray, or evanescent, fields generated by macroscopic structures at or near the surface of a sample, and as such cannot reveal features on the atomic scale. Chapter 12 discusses magnetic force microscopy, where surface topography and domain structure on magnetic media are imaged. As in Chapter 11, since magnetostatics is a cooperative phenomena, no atomically resolved images are feasible. Chapter 13, which deals with atomic force microscopy, differs from the other two microscopies in that it can image both macroscopic and microscopic features related to the atomic structure of surfaces. Accordingly, the chapter presents both microscopic and macroscopic interactions between the atoms at the apex of the force-sensing tip and those at and near the surface of a sample. Following is a discussion of the principles of operation of atomic force microscopy, the tip-sample approach mechanism, a comparison between atomic, electrostatics and magnetostatic interactions that can exist simultaneously, and a summary of key accomplishments. The photographs in the book are presented in an effort to demonstrate some of the state-of-the-art accomplishments of scanning force microscopy. Because of time constraints and the proprietary nature of some of the technologies, it was decided not to include in this book mechanical isolation systems, fabrication of levers and tips, analog and digital electronic feedback and image processing, and design of tripod and tube piezoelectric scanners and their drivers. Gray-scale photographs were kindly provided by Alvarado et al., Binnig et al., Elings et al., Hansma et al., Wickramasinghe et al., Rugar et al., Quate et al., Park et al., and Landman et al. Also, G. Binnig, C. F. Quate, Ch. Gerber, T. Albrecht, S. Chang, P. K. Hansma, P. Grütter, U. Landman, H. J. Mamin, R. McAllister, G. M. McClelland, D. H. Pohl, D. Rugar, B. D. Terris, and H. K. Wickramasinghe provided an abundance of advice.

The author would like to acknowledge contributions of his colleagues, L. S. Bell, M. J. Gallagher, T. D. Henson, S. C. Howells, D. A. Iams, J. T. Ingle, R. Porter, and L. Yi. Special thanks go to V. Elings, D. Bocek, and the rest of the gang at Digital Instruments who made many things possible. The support of R. G. Brandt (ONR), A. Harvey (NSF), H. A. Jenkinson (USAAMCCOM), R. Miceli (ONR), H. R. Schlossberg (AFOSR), R. Schmulian (IBM), and H. K. Wickramasinghe (IBM) are gratefully acknowledged. Editorial, art work, and camera-ready typesetting benefitted from M. Whitney, M. Wright, L. Schadler, M. Dorsey, and M. Sargent's technical word processing software, *PS*[c]. Clearly, mistakes, omission of factors of 2 or π, and missing references are all my responsibility.

Contents

PART TWO. SCANNING FORCE MICROSCOPES

PART THREE. SCANNING FORCE MICROSCOPY

Scanning Force Microscopy

REVISED EDITION

1
Mechanical Properties of Levers

1.1. Introduction

The heart of a scanning force microscope is a sharp tip that interacts with a force at the surface of a sample. The tip, therefore, must have certain material properties, such as conductivity and permeability, that determine the strength of the interaction. The tip is mounted on a flexible beam whose geometrical and material properties make it possible to probe the force with a high sensitivity. The interaction of the tip with the force is sensed by the resultant deflection of the beam on which the tip is mounted. The role of the beam is to translate the force acting on the tip into a deflection that subsequently can be monitored by various means. Among these, tunneling of electrons, capacitance, optical interferometry, optical polarization, and optical deflection have recently been developed to a high degree of sophistication. To understand the particular properties required of the beam, we outline in this chapter the theory that models deformation and resonance frequencies of the beam as a function of material and geometrical parameters (Shigley 1963, Harris 1988). The discussion will pertain to mechanical beams that are supported at one end, and are therefore called cantilevers, or in short, levers. The forces acting on such a lever are: (1) its weight, which is a distributed force; (2) the weight of a tip attached at the free end, which is a concentrated force; and (3) a driving force acting on the lever that acts as a concentrated force. The number of degrees of freedom of the lever will equal the number of its natural frequencies of vibration. In particular, since the lever has a distributed weight, it can vibrate at an infinite number of frequencies. In reality, however, the lever will vibrate at a few or at only one frequency, where each frequency has a deformation profile, called the normal mode of vibration. Each point along the length of the lever can oscillate with a harmonic motion, except for the support and node points, which are stationary. For the fundamental vibration frequency, the only point that is stationary is the support at one end of the lever. Clearly, if the lever vibrates at more than one frequency, it will assume a shape composed of a linear combination of its normal modes, provided the amplitude of vibration is small enough.

Two methods used to solve for the frequencies and normal modes of a vibrating lever are the Rayleigh and classical methods. Rayleigh's method uses the conservation of strain and kinetic energy for finding an approximate value for the fundamental vibration frequency. To that end, one calculates the strain energy at the point where the lever reaches its maximum deflection, and equates this energy to the kinetic energy of the lever at the point of zero deformation. The classical method uses Newton's second law of motion, which yields, for a finite number of degrees of freedom, a set of second-order differential equations. For a system having a distributed parameter, however, one gets a set of partial differential equations. For both methods, however, the set of equations is solved by invoking the appropriate boundary conditions, which in our case are dictated by the geometry of the lever. We will use both of these methods to evaluate the properties of solid rectangular and cylindrical levers.

1.2. Stress and Strain

Let us consider a three-dimensional body in a rectangular coordinate system and a point inside this body on which a force **F** acts. One can define three stress components associated with an element of area A_x perpendicular to the x direction. The first component is the normal stress σ_{xx} given by

$$\sigma_{xx} = \lim_{A_x \to 0} \frac{F_x}{A_x} , \tag{1.1}$$

and the other two components are the shear stresses τ_{xy} and τ_{xz} given by

$$\tau_{xy} = \lim_{A_x \to 0} \frac{F_y}{A_x} \tag{1.2}$$

and

$$\tau_{xz} = \lim_{A_x \to 0} \frac{F_z}{A_x} . \tag{1.3}$$

In a similar manner, we can define the three stress components associated with elements of area perpendicular to the z directions, as shown in Fig. 1.1.

Consequently, a vectorial field of force acting on all points in the body gives rise to a tensorial field of stress that can be expressed at each point in the body by the nine components σ_{ij}, where i and j are the values of x, y, and z, respectively. For simplicity, we will drop the tensorial notation. Now, a body acted on by an external force will experience a deformation that can be described by either of two types

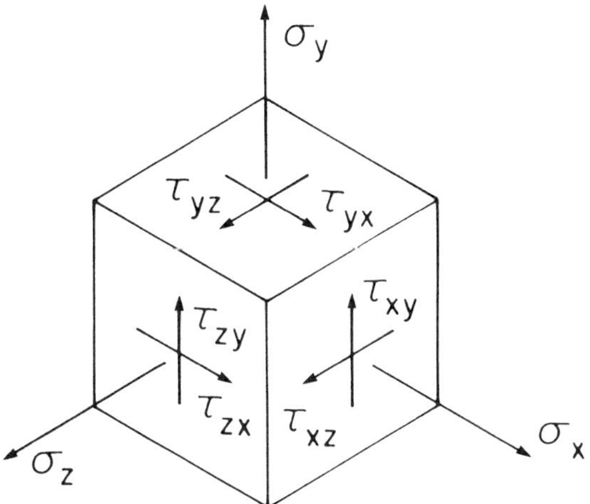

Fig. 1.1 The three stress components σ_x, σ_y, and σ_z; and the six strain components $\tau_{i,j}$ where i and j denote the x, y, and z directions.

of strain tensors. The first type of strain describes the relative elongation δ of a bar of length ℓ, and is defined by

$$\epsilon = \delta/\ell \ . \tag{1.4}$$

Here ϵ, the strain, is in the same direction as the stress. As long as the deformed body is in its elastic regime, there will be a linear relationship between the stress and the strain at any given point in the body, that can be expressed by

$$\sigma = E\epsilon \ . \tag{1.5}$$

The constant E is the modulus of elasticity, or Young's modulus, and is expressed in units of N/m². The second kind of strain, denoted by γ, is the relative deformation in a direction perpendicular to the direction of the stress, and is given by

$$\tau = G\gamma \ . \tag{1.6}$$

The constant G is the shear modulus of elasticity, or modulus of rigidity, and is also expressed in units of N/m². The constants E and G, therefore, describe the resistance of the lever to bending. Using Eqs. (1.1), (1.4), and (1.5) gives

$$\delta = \frac{\ell}{AE} F \,, \tag{1.7}$$

which relates the force to the elongation of a spring.

1.3. Moments

Consider now Fig. 1.2, which shows a bent body with its associated parameters. The bending about the x axis gives rise to a compression above a line passing through the center of the beam, and a dilation below that line. This line, along which there is no compression or dilation, is called the neutral axis, and the distance z we use in the following is measured relative to this axis. We note that the angle of bending, $d\phi$, is related to the distance ds by

$$R = ds/d\phi \,, \tag{1.8}$$

where R is the radius of bending. The deformation (compression or dilation) of a "fiber" shown as a dashed line, at a distance z from the neutral axis, is given by

$$dy = zd\phi \,, \tag{1.9}$$

which yields its strain

$$\epsilon = -\frac{dy}{ds} = -\frac{z}{R} \,. \tag{1.10}$$

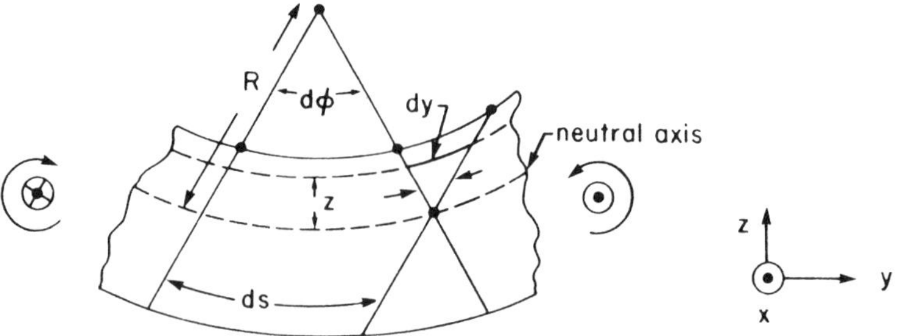

Fig. 1.2 In a bent body, the bending about the x axis gives rise to a compression above the neutral axis and a dilation below that axis.

The stress is then

$$\sigma = -E\frac{z}{R}. \tag{1.11}$$

Consequently, both the stress and the strain are proportional to the distance z from the neutral axis. At equilibrium, the external moment M will equal the internal moments generated by the stress that are distributed throughout any given cross section A. Using the definitions of the area moment of inertia given by

$$I = \int_A z^2 dA , \tag{1.12}$$

and the moment given by

$$M = \int_A z\sigma dA , \tag{1.13}$$

we get the important result

$$\frac{1}{R} = \frac{M}{EI}. \tag{1.14}$$

1.4. Spring Constant

A general equation that yields solutions for the various aspects of a deflected beam, such as its deflection and resonance frequency, can now be written. Using the expression for the curvature of a plane curve,

$$\frac{1}{R} = \frac{d^2z/dy^2}{[1 + (dz/dy)^2]^{3/2}}, \tag{1.15}$$

in Eq. (1.14) we get

$$\frac{\partial^2 z}{\partial y^2} = \frac{M}{EI}. \tag{1.16}$$

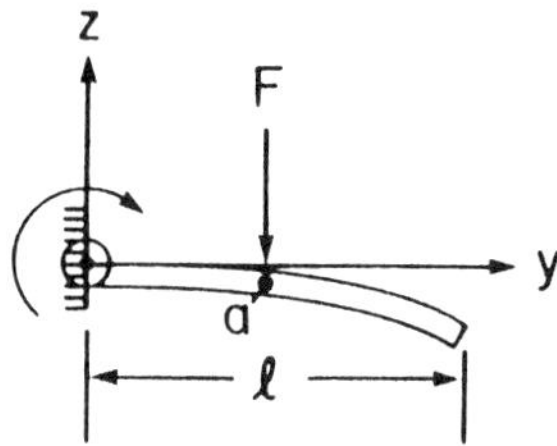

Fig. 1.3 A lever anchored rigidly on its left side, the force **F** acting on its right side downward, and the resulting moment **M**.

Here we neglected the derivative term in the denominator, which is justified for small angles. Equation (1.16), which was obtained by equating the curvature of the lever to the bending moment at each point along the lever, is the equation of motion from which the classical solutions to the problem of a vibrating lever will be derived later. Figure 1.3 shows a lever anchored rigidly on its left side and a force F acting on its right side downward. We find that the moment at any given point $y \leq$ a, produced by a force acting at point a, is

$$M = F(y - \mathrm{a}) \ . \tag{1.17}$$

Inserting the value of M in Eq. (1.16) and integrating twice gives the deflection of the lever

$$z = \frac{Fy^2}{6EI}(y - 3a) \ , \tag{1.18}$$

and setting $y = a = \ell$ gives

$$z = - \frac{\ell^3}{3EI} F \ . \tag{1.19}$$

We can now define a spring constant (compliance) k by

$$k = \left| \frac{F}{z} \right| , \tag{1.20}$$

and get

$$k = 3\,\frac{EI}{\ell^3}\,. \tag{1.21}$$

We assumed that the bending is small enough to stay within the range of elasticity so that F is not a function of z. To calculate the spring constant of a given body, we need to know its Young's modulus E, area moment of inertia I, and length ℓ. It is useful at this point to note that n levers arranged in series yield a total deflection z given by

$$z = F \sum_n \frac{1}{k_n}\,, \tag{1.22}$$

and the effective spring constant of such a system of levers is given by

$$k^{-1} = \sum_n \frac{1}{k_n}\,. \tag{1.23}$$

Equation (1.23) tells us that the longer the lever is, the smaller its spring constant will be. However, if the levers are arranged in parallel, then

$$F = \sum_n k_n z\,, \tag{1.24}$$

and the effective spring constant of the system is given by

$$k = \sum_n k_n\,. \tag{1.25}$$

We find, therefore, that the wider the lever is, the larger its spring constant will be.

1.5. The Rayleigh Solution to a Vibrating Lever

The fundamental resonance frequency of a lever can be computed approximately by equating its strain energy, when the deformation is at a maximum, to the kinetic energy when the deformation is zero. A more accurate method, which gives all the resonance frequencies together with the modes of vibration, can be obtained by solving the equation of motion subject to the boundary conditions. We will start with the first case, and consider a vibrating lever whose strain energy is given by

$$W_s = \frac{EI}{2} \int_0^{\ell} \left[\frac{\partial^2 z}{\partial y^2} \right]^2 dy \,. \tag{1.26}$$

Here ℓ is the length of the lever and $z = z(y)$ describes the lateral position of each point along the neutral axis. As an example, by using Eq. (1.16), we find that the static deflection of a lever, in terms of its maximum deflection z_0, is

$$z(y) = \frac{z_0}{3\ell^4} [y^4 - 4y^3\ell + 6y^2\ell^2] \,. \tag{1.27}$$

This equation can be derived by considering the boundary conditions $z(\ell) = z_0$, $z(0) = \partial z(0)/\partial y = 0$, and $\partial^2 z(\ell)/\partial y^2 = 0$. The second derivative of $z(y)$ is

$$\frac{\partial^2 z}{\partial y^2} = \frac{4z_0}{\ell^4} (y - \ell)^2 \,, \tag{1.28}$$

and the maximum strain energy W_s is therefore

$$W_s = \frac{8}{5} \, \mathrm{EI} \, \frac{z_0^{\,2}}{\ell^3} \,. \tag{1.29}$$

The kinetic energy of the lever, W_k, is given by

$$W_k = \int_0^{\ell} \frac{A\rho}{2} \left[\frac{\partial z}{\partial t} \right]^2 dy \,, \tag{1.30}$$

where ρ is the mass density. Denoting by ω the angular frequency of vibration, we find that the maximum kinetic energy for a given mode is

$$W_{k,max} = \frac{A\rho\omega^2}{2} \int_0^{\ell} z^2(y) \, dy \,, \tag{1.31}$$

which occurs when the deflection is zero. Using Eq. (1.27) in Eq. (1.31) yields

$$W_{k,max} = \frac{52}{405} A\rho\omega^2 \ell z_0{}^2 \, . \tag{1.32}$$

By equating the strain and kinetic energies, Eqs. (1.29) and (1.32), we get

$$\omega = \kappa^2 \sqrt{\frac{EI}{A\rho}} \, , \tag{1.33}$$

with

$$\kappa = 1.8788/\ell \, . \tag{1.34}$$

This approximate value of κ can be compared to the accurate value, $\kappa = 1.875/\ell$, calculated in the next section.

1.6. The Classical Solution to a Vibrating Lever

The classical solution, which is the more exact one, produces all the resonance frequencies of the lever together with the modes of vibration. In this method we first derive the equation of motion and then solve it using the boundary conditions of the lever. We start by defining the shear force V by

$$V = \frac{\partial M}{\partial y} \, , \tag{1.35}$$

and then equate dV, acting on an element of mass $\rho A dy$, with the acceleration, obtaining

$$dV = -\rho A dy \, \frac{\partial^2 z}{\partial t^2} \, . \tag{1.36}$$

Combining Eqs. (1.16) and (1.36) and assuming that EI is a constant yields the equation of motion of the lever

$$\mathrm{EI} \, \frac{\partial^4 z}{\partial y^4} + \rho A \, \frac{\partial^2 z}{\partial t^2} = 0 \, , \tag{1.37}$$

whose solution is

$$z(y,t) = z(y)[\cos(\omega_n t + \theta)] , \tag{1.38}$$

where n is the order of the mode. By defining a parameter κ such that

$$\kappa^4{}_n = \frac{\omega^2{}_n \rho A}{EI} , \tag{1.39}$$

we can reduce Eq. (1.37) to the form

$$\frac{d^4 z(y)}{dy^4} = \kappa^4 z . \tag{1.40}$$

1.7. Normal Modes

The solution to the equation of motion, Eq. (1.40), is

$$z(y) = A_1 \sin \kappa y + A_2 \cos \kappa y + A_3 \sinh \kappa y + A_4 \cosh \kappa y , \tag{1.41}$$

which, for convenience when applying boundary conditions, can also be written as

$$z(y) = A'(\cos \kappa y + \cosh \kappa y) + B'(\cos \kappa y - \cosh \kappa y)$$
$$+ C'(\sin \kappa y + \sinh \kappa y) + D'(\sin \kappa y - \sinh \kappa y) . \tag{1.42}$$

The deflection, slope, moment, and shear force of the deformed lever are proportional to z, $\partial z/\partial y$, $\partial^2 z/\partial y^2$, and $\partial^3 z/\partial y^3$, respectively. These derivatives are given explicitly by

$$\frac{\partial z}{\partial y} = \kappa[A'(-\sin \kappa y + \sinh y) - B'(\sin \kappa y + \sinh \kappa y)$$
$$+ C'(\cos \kappa y + \cosh \kappa y) + D'(\cos \kappa y - \cosh \kappa y)] , \tag{1.43}$$

$$\frac{\partial^2 z}{\partial y^2} = \kappa^2[A'(-\cos \kappa y + \cosh \kappa y) - B'(\cos \kappa y + \cosh \kappa y)$$
$$+ C'(-\sin \kappa y + \sinh \kappa y) - D'(\sin \kappa y + \sinh \kappa y)] , \tag{1.44}$$

and

$$\frac{\partial^3 z}{\partial y^3} = \kappa^3[A'(\sin \kappa y + \sinh \kappa y) + B'(\sin \kappa y - \sinh \kappa y)$$

$$+ C'(-\cos \kappa y + \cosh \kappa y) - D'(\cos \kappa y + \cosh \kappa y)] \,. \tag{1.45}$$

The boundary conditions for the oscillating lever give $z = \partial z/\partial y = 0$ at $y = 0$ so that $A' = C' = 0$. At $y = \ell$ we have $\partial^2 z/\partial y^2 = \partial^3 z/\partial y^3 = 0$, which yields

$$0 = B'(\cos \kappa\ell + \cosh \kappa\ell) + D'(\sin \kappa\ell + \sinh \kappa\ell) \tag{1.46}$$

and

$$0 = B'(\sin \kappa\ell - \sinh \kappa\ell) + D'(-\cos \kappa\ell - \cosh \kappa\ell) \,. \tag{1.47}$$

Equations (1.46) and (1.47) give

$$\frac{D'}{B'} = -\frac{\cos \kappa\ell + \cosh \kappa\ell}{\sin \kappa\ell + \sinh \kappa\ell} = \frac{\sin \kappa\ell - \sinh \kappa\ell}{\cos \kappa\ell + \cosh \kappa\ell} \,, \tag{1.48}$$

which reduces to the simple implicit equation

$$\cos \kappa\ell \cosh \kappa\ell + 1 = 0 \,. \tag{1.49}$$

The first five values of $\kappa_n \ell$ and D'/B' are

$$\kappa_n \ell = 1.875,\ 4.694,\ 7.855,\ 10.996,\ 14.137 \tag{1.50}$$

and

$$\frac{D'}{B'} = -0.7341,\ -1.0185,\ -0.9992,\ -1,\ -1 \,. \tag{1.51}$$

For higher values one can use the approximations

$$\kappa_n \ell \simeq \frac{2n-1}{2}\pi \tag{1.52}$$

and

$$\frac{D'}{B'} = -1 \,. \tag{1.53}$$

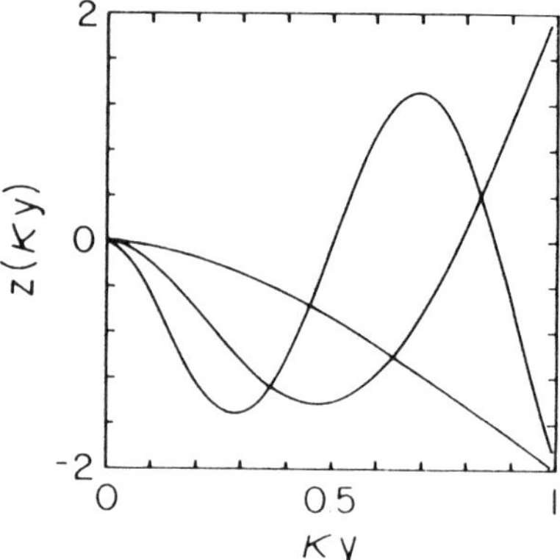

Fig. 1.4 The first three modes of vibration of a lever. Note that the curves do not resemble sine waves because the mass is distributed along the lever.

The frequencies of vibration are therefore given by

$$\omega_n = \kappa^2{}_n \sqrt{\frac{EI}{\rho A}} \ . \tag{1.54}$$

The normal mode for each frequency can be calculated by using the values of $\kappa_n \ell$ and D'/B', yielding

$$z(y,t) = A'_0 \left[(\cos \kappa y - \cosh \kappa y) + \frac{D'}{B'}(\sin \kappa y - \sinh \kappa y) \right] \cos(\omega_1 t + \theta) \ , \tag{1.55}$$

where A'_0 is determined by the amplitude of vibration. Figure 1.4 shows the first three modes of a vibrating lever where the curves do not resemble sine waves because the mass is distributed along the lever.

1.8. Lumped Systems

Consider now the behavior of a lumped system that consists of both a concentrated and a distributed mass, m_c and m_d, respectively. For the fundamental mode and for $m_c = 0$,

$$\omega = \sqrt{\frac{\kappa^4 EI}{\rho A}} \ . \tag{1.56}$$

Using Eq. (1.21) yields

$$\omega = \sqrt{\frac{k}{0.24m_d}} \; . \tag{1.57}$$

We can now define an effective mass by

$$m_{eff} = m_c + 0.24m_d \; , \tag{1.58}$$

and obtain the general expression for the frequency of the fundamental mode of a lever having both distributed and concentrated masses

$$\omega = \sqrt{\frac{k}{m_{eff}}} \; . \tag{1.59}$$

Equation (1.59) is the basic equation that we will use to calculate the resonance frequency of a lumped system consisting of either a rectangular or a solid cylindrical lever.

1.9. Examples

The area moment of inertia of a rectangular lever, using Eq. (1.12), is

$$I = \frac{wt^3}{12} \; , \tag{1.60}$$

where w is the width and t is the thickness of the lever, which are perpendicular and parallel to the direction of the bending, respectively. Using this value of I in Eq. (1.21) yields the spring constant of the rectangular lever

$$k = \frac{Ewt^3}{4\ell^3} \; . \tag{1.61}$$

The area moment of inertia of a solid cylindrical lever is

$$I = \frac{\pi r^4}{4} \; , \tag{1.62}$$

where r is the radius of the lever from which we obtain its spring constant

$$k = \frac{3\pi E r^4}{4\ell^3} . \quad (1.63)$$

Note that the geometry of the lever determines I, while the material properties determine E and ρ. The resonance frequency of a rectangular lever, with and without a concentrated load, using Eq. (1.57), is

$$\omega = \sqrt{\frac{E w t^3}{4\ell^3(m_c + 0.24 w t \ell \rho)}} \quad (1.64)$$

and

$$\omega \simeq \frac{t}{\ell^2}\sqrt{\frac{E}{\rho}} , \quad (1.65)$$

respectively. Likewise, the resonance frequency of a solid cylindrical lever, with and without a concentrated load, is

$$\omega = \sqrt{\frac{3\pi E r^4}{4\ell^3(m_c + 0.24\pi r^2 \ell \rho)}} \quad (1.66)$$

and

$$\omega \simeq \sqrt{3}\,\frac{r}{\ell^2}\sqrt{\frac{E}{\rho}} , \quad (1.67)$$

respectively. We have calculated the resonance frequency and deflection of rectangular and solid cylindrical levers in terms of their material parameters E and ρ. Conversely, we can derive the material parameters E and ρ if we have the geometry of a lever, its resonance frequency, and its spring constant. For a rectangular lever we get

$$E = \frac{4k}{w}\,\frac{\ell^3}{t^3} \quad (1.68)$$

and

$$\rho = \frac{k}{0.24\pi^2 f^2 \ell w t} . \quad (1.69)$$

For a solid cylindrical lever we get

$$E = \frac{4k}{3\pi r}\,\frac{\ell^3}{r^3} \tag{1.70}$$

and

$$\rho = \frac{k}{0.24\pi^3 f^2 \ell r^2}\,. \tag{1.71}$$

To calculate the resonance frequency of a lever we have to know (1) its geometry, (2) its mass distribution, (3) the concentrated mass, and (4) modulus of elasticity. For a solid cylindrical lever, we can define two coefficients

$$c_1 = \frac{1}{2\pi}\left[\frac{3E}{\rho}\right]^{1/2} \tag{1.72}$$

and

$$c_2 = \frac{3\pi}{4}E\,, \tag{1.73}$$

and get any two of the four parameters ℓ, k, ν, and r from the two others, as shown in Table 1.1. These relationships will be useful when designing levers according to given specifications. Table 1.2 gives representative values of ρ and E for several materials of interest, which will assist in designing proper levers for scanning force microscopes.

Table 1.1 The relationships between ℓ, r, k, and ν for a cylindrical lever.

$(\ell,r) \rightarrow (\nu,k)$:	$\nu = c_1 r/\ell^2$	$k = c_2 r^4/\ell^3$
$(\ell,k) \rightarrow (r,\nu)$:	$r = (k\ell^3/c_2)^{1/4}$	$\nu = c_1 r/\ell^2$
$(\ell,\nu) \rightarrow (r,k)$:	$r = \nu\ell^2/c_1$	$k = c_2 r^4/\ell^3$
$(r,k) \rightarrow (\ell,\nu)$:	$\ell = (c_2 r^4/k)^{1/3}$	$\nu = c_1 r/\ell^2$
$(r,\nu) \rightarrow (\ell,k)$:	$\ell = (c_1 r/\nu)^{1/2}$	$k = c_2 r^4/\ell^3$
$(k,\nu) \rightarrow (\ell,r)$:	$\ell = (kc_1{}^4/c_2\nu^4)^{1/5}$	$r = \nu\ell^2/c_1$

Table 1.2 Mass density ρ and modulus of elasticity E.

	$\rho(\mathrm{kg/m^3})$	$E(10^{11}\mathrm{N/m^2})$
graphite	2,250	2.0
Si	2,330	1.79
SiO_2	2,200	0.6
Si_3N_4	3,100	1.5
Al	2,700	0.7
Ti	4,500	1.16
Fe	7,870	1.93
Ni	8,900	2.07
Mg	11,740	0.41
W	19,300	3.45
Cu	8,960	1.1
Pt	24,200	1.7
Au	19,300	0.8
Ir	2,250	5.2

Figure 1.5 shows a cylindrical Ni lever with a diameter of 12.5 μm whose end was electrochemically etched to roughly 1000 Å. The lever, used for magnetic force microscopy, has a resonance frequency of 9 kHz and a spring constant of 1 N/m.

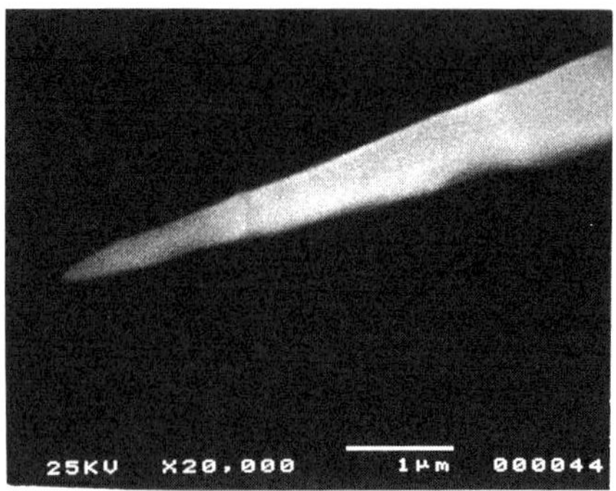

Fig. 1.5 An alumel lever used for magnetic force microscopy.

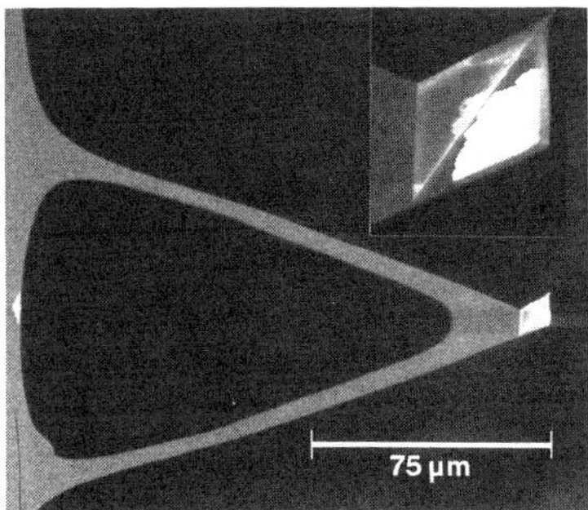

Fig. 1.6 A single crystal silicon tip integrated with an amorphous silicon nitride lever. (Courtesy: S. Akamine, R. C. Barrett, and C. F. Quate, Stanford University.)

Figure 1.6 shows a microfabricated Si_3N_4 lever with its integrated pyramidal tip used for atomic force microscopy. A rectangular lever is typically 100 μm long, 10 μm wide, and 0.6 μm thick, has a resonance frequency of 66 kHz and a spring constant 0.08 N/m. For a detailed description of microfabricated AFM levers, see Albrecht et al. (1990a,b), Akamine et al. (1990a,b), Grütter et al. (1990c), Drexler (1990), Wolter et al. (1990), Wise and Orr (1990), Schmidt et al. (1990), Hellemans et al. (1990), Buser et al. (1992), Grigg et al. (1992a), Ximen and Russell (1992), Keller et al. (1992a), and Kado et al. (1992).

1.10. Summary

The main result obtained in this chapter gives the resonance frequencies of a solid rectangular lever

$$\omega = \sqrt{\frac{Ewt^3}{4\ell^3(m_c + 0.24wt\ell\rho)}}$$

and solid cylindrical lever

$$\omega = \sqrt{\frac{3\pi Er^4}{4\ell^3(m_c + 0.24\pi r^2\ell\rho)}} ,$$

where E, m_c, and ρ are the modulus of elasticity, concentrated mass, and mass density of the lever, and w, t, r, and ℓ are the width, thickness, radius, and length of the lever, respectively.

2
Resonance Enhancement

2.1. Introduction

This chapter presents the theory of interaction of a tip mounted on a vibrating lever and an inhomogeneous external force where the mechanical resonance of the lever plays a major role in the response of the system. We start by describing the basic equations governing the operation of a bimorph used to position and vibrate the lever. Next, we introduce the concept of an effective spring constant, which accounts for force derivatives acting on a sharp tip attached to the lever. We then solve the equation of forced motion of the damped lever for three cases. In the first case, the lever, which is mounted on a vibrating bimorph, interacts with a force derivative, and its amplitude of vibration is monitored by a probe. In the second case, the lever is made to vibrate by means of the force derivative belonging to a vibrating sample and its amplitude is monitored by a probe. In the third case, the force derivative, belonging to a vibrating tip of a scanning tunneling microscope, induces a vibration in the lever whose amplitude is monitored by the tunneling current. We plot the response of each system as a function of the exciting frequency and as a function of the force derivative and discuss the results. For clarity, we use in the examples the parameters $a = 1$, $k = 1$ N/m, and $Q = 10$ where a, k, and Q are the amplitude of vibration of the bimorph, the spring constant of the lever, and its quality factor, respectively, and the operating frequency ω is scaled to that of the resonance frequency of the free lever. Also, the frequency plots have a force derivative equal to 0.5 N/m, while the force derivative plots have a frequency equal to the resonance frequency of the free lever, and the curves showing the derivatives of the amplitude are arbitrarily scaled.

2.2. Bimorph Driver

A bimorph consists of two layers of a piezoelectric polycrystalline ceramic bonded to a thin metal shim sandwiched in the middle. The bimorph will bend when a voltage is impressed on its two electrodes, and can therefore be used to adjust the position of a lever attached to it as well as vibrate it. The deflection Δz, resonance frequency f_r, and capacitance C of a rectangular bimorph, shown in Fig. 2.1 (Piezo Electric Products Inc.; see also, Chen 1992a,b), are

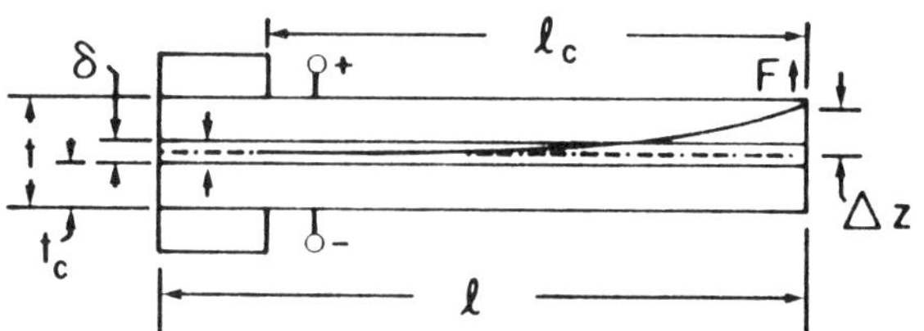

Fig. 2.1 The geometry of a rectangular bimorph.

$$\Delta z = 3\beta d_{31} \left[\frac{\ell_c}{t}\right]^2 \left[1 + \frac{\delta}{t}\right] t_c E \,, \tag{2.1}$$

$$f_r = \frac{0.16t}{\ell_c{}^2} \left[\frac{E_{11}}{\rho}\right]^{1/2} , \tag{2.2}$$

and

$$C = \epsilon_r \frac{\epsilon_0 \ell w}{2t_c} \,. \tag{2.3}$$

Here, ℓ, t, and w are the bimorph length, thickness, and width, respectively, ℓ_c is its free part, δ is the thickness of center shim and adhesive, t_c is the thickness of a single layer of the piezoceramic, $d_{31} = \sigma/E_{11}$ is the transverse strain coefficient, $E = V/2t_c$ is the electric field, V is the voltage, E_{11} is Young's modulus, ϵ_0 and ϵ_r are the dielectric constant of free space and its relative value for the bimorph, respectively, ρ is the average density, and $\beta \simeq 1$ (for $V < 40$ V/mm) is a nonlinear constant. For a typical bimorph with thickness of 0.5 mm, a length of 10 mm, and a piezoelectric strain coefficient of 180×10^{-12}, the displacement is approximately 4 nm/V. It should be mentioned in passing that Anders and Heiden (1988, 1990), Göddenhenrich et al. (1988), Tansock and Williams (1992), and Tortonese et al. (1993) have used a bimorph for sensing the force acting between tip and sample. The amplitude of vibration of a lever attached to the vibrating bimorph, however, will be larger than that of the bimorph if the frequency of vibration is on or close to its resonance frequency. The amplitude of vibration of the lever in this case will be determined by its damping, or quality factor, Q, which can be measured from the mechanical frequency response. Q is given by the ratio of the resonance frequency ω_0 and the full bandwidth, at 0.707 of the maximum amplitude,

$$Q = \frac{\omega_0}{\Delta\omega} . \tag{2.4}$$

For further discussions see, for example, Tabib-Azar (1990), Whitehouse (1990), and Blom et al. (1992).

2.3. Effective Spring Constant

The spring constant was defined by $k = |F/z|$, where F is the force acting on the spring and z the resultant deflection. Another definition of the spring constant k can be derived from the potential energy W of a deformed spring, or lever,

$$W = \frac{1}{2} kz^2 , \tag{2.5}$$

by taking the second derivative of the energy W in respect to z,

$$k = \frac{\partial^2 W}{\partial z^2} . \tag{2.6}$$

Equation (2.6) is convenient for finding the effective spring constant in the presence of a force $F(z)$ that has a derivative in the direction of deflection of the spring. In this case we can expand the force to first order,

$$F(z) = F(z_0) + \frac{\partial F(z_0)}{\partial z} \delta z , \tag{2.7}$$

and get an effective spring constant, k',

$$k' = k - F_1 , \tag{2.8}$$

where $F_1 = \partial F/\partial z$. Usually the probe that senses the vibration of the tip is above the tip while the sample is below the tip. For convenience, we denote a force derivative pointing toward the sample as positive. Therefore, a tip-sample attractive force with a positive derivative decreases the resonance frequency of a lever.

2.4. Bimorph-Driven Lever

Figure 2.2 shows a tip mounted on a lever that is attached to a bimorph vibrating with an amplitude a and frequency ω. It has been shown recently that instead of using a bimorph, one can use the photothermal effect and drive the lever with a modulated laser beam. Now, the positions of the bimorph and the lever are given by

$$u = u_0 + a \exp(i\omega t) \tag{2.9}$$

and

$$z = z_0 + \zeta , \tag{2.10}$$

respectively, the position of the sample is given by

$$g = 0 , \tag{2.11}$$

and ζ is yet an unknown function. Here BM is the bimorph, S is the sample, ℓ and t are the lever and the force-sensing tip, and P is the tip position probe.

We wish to find the amplitudes of vibration of the lever and the phase angle θ between the vibration of the bimorph and the lever for two cases. In the first case, the interaction force F acting between the tip at the end of the lever and the sample is constant, and in the second case, it has a derivative along the direction of vibration of the lever.

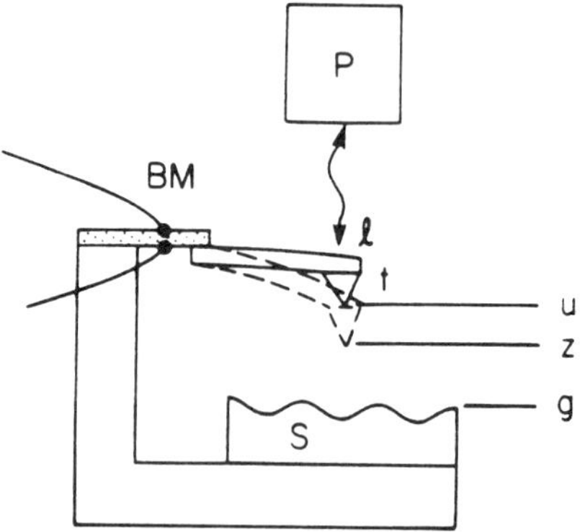

Fig. 2.2 The geometry of the bimorph-driven lever where u and z are the positions of the undeflected and deflected tip, respectively, and g is the position of the sample.

2.4.1. Constant Interaction Force

Let us first consider the case where the interaction force is uniform, namely,

$$F = F_0 \,. \tag{2.12}$$

The equation of motion of the lever,

$$m \frac{\partial^2 z}{\partial t^2} + \gamma \frac{\partial z}{\partial t} + k(z - u) = F_0 \,, \tag{2.13}$$

has three independent parameters: the effective mass of the lever m, the spring constant of the lever k, and the dissipation term γ. Inserting Eqs. (2.9) and (2.10) in the equation of motion of the lever gives

$$m \frac{\partial^2 \varsigma}{\partial t^2} + \gamma \frac{\partial \varsigma}{\partial t} + k[z_0 + \varsigma - u_0 - a\exp(i\omega t)] = F_0 \,. \tag{2.14}$$

Since at equilibrium the sample-lever force equals the restoring force of the lever

$$F_0 = k(z_0 - u_0) \,, \tag{2.15}$$

we get

$$m \frac{\partial^2 \varsigma}{\partial t^2} + \gamma \frac{\partial \varsigma}{\partial t} + k[\varsigma - a\exp(i\omega t)] = 0 \,. \tag{2.16}$$

The vibration of the lever will have the same frequency as that of the bimorph and can be written as

$$\varsigma = A_b \exp[i(\omega t - \theta)] \,, \tag{2.17}$$

where θ is a time-independent phase angle. Using Eq. (2.17) in Eq. (2.16) yields

$$A_b(\omega)[k - \omega^2 m + i\omega\gamma] = ak\exp(i\theta) \,. \tag{2.18}$$

Isolating the amplitude of vibration of the lever by taking the absolute value of Eq. (2.18) gives

$$A_b(\omega) = \frac{ak}{[(k - \omega^2 m)^2 + \omega^2\gamma^2]^{1/2}} \,. \tag{2.19}$$

Here we can express the dissipation term as

$$\gamma = m\omega_0/Q \,, \tag{2.20}$$

where

$$\omega_0 = \sqrt{k/m} \tag{2.21}$$

is the resonance frequency of the free lever for $Q >> 1$. It is helpful to define a bimorph-driven lever response function, $G_b(\omega)$, by

$$G_b(\omega) = \frac{A_b(\omega)}{a} \,, \tag{2.22}$$

where

$$G_b(\omega) = \frac{Q}{[Q^2(1 - \omega^2/\omega^2{}_0)^2 + \omega^2/\omega^2{}_0]^{1/2}} \,. \tag{2.23}$$

Decomposing Eq. (2.18) into imaginary and real components gives for the sine and cosine functions

$$\sin\theta_b = G_b(\omega)\,\frac{\omega}{Q\omega_0} \tag{2.24}$$

and

$$\cos\theta_b = G_b(\omega)\,\frac{\omega_0{}^2 - \omega^2}{\omega_0{}^2} \,. \tag{2.25}$$

On resonance, where $\omega = \omega_0$, we get

$$G_{b\,0} = Q \,, \tag{2.26}$$

$$A_{b\,0} = aQ \,, \tag{2.27}$$

and

$$\theta_{b\,0} = \frac{\pi}{2} \,. \tag{2.28}$$

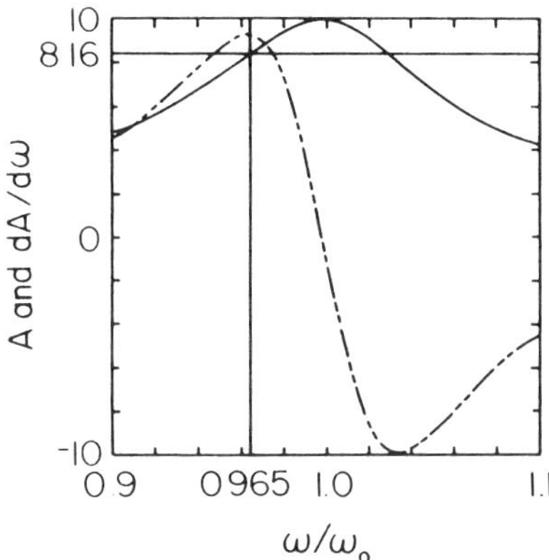

Fig. 2.3 $A_b(\omega)$ (full line) and $\partial A_b/\partial\omega$ (dashed line) are the amplitude of vibration and the derivative as a function of ω/ω_0 for $F_1 = 0$. Note that the derivative obtains its maximum value slightly off resonance.

Figure 2.3 shows $A_b(\omega)$ (full line) and $\partial A_b/\partial\omega$ (dashed line), respectively, as a function of ω/ω_0, where $A_b(\omega)$ and $\partial A_b(\omega)/\partial\omega$ obtain their extrema at ω_0 and $\omega_0 \pm \Delta\omega$, respectively, and the amplitude on resonance is $Q = 10$. We will later discuss the importance of operating the system at $\omega = \omega_0 \pm \Delta\omega$, where the derivative obtains its maximum value.

2.4.2. Nonuniform Interaction Force

In this case, Eqs. (2.9) through (2.11) still hold, but the interaction force and the equation of motion become

$$F(z) = F_0(z_0) + F_1(z_0)\zeta \tag{2.29}$$

and

$$m\frac{\partial^2 z}{\partial t^2} + \gamma\frac{\partial z}{\partial t} + k(z - u) = F(z)\ , \tag{2.30}$$

respectively. Note that we expanded the nonuniform force to first order in the perturbation term ζ around $z = z_0$. The equivalent of Eq. (2.14) is now

$$m \frac{\partial^2 \zeta}{\partial t^2} + \gamma \frac{\partial \zeta}{\partial t} + k[z_0 + \zeta - u_0 - a\exp(i\omega t)] = F_0 + F_1 \zeta \ , \qquad (2.31)$$

and, since Eq. (2.15) still holds, we get for Eq. (2.31)

$$m \frac{\partial^2 \zeta}{\partial t^2} + \gamma \frac{\partial \zeta}{\partial t} + k[\zeta - \mathrm{a}\exp(i\omega t)] = F_1 \zeta \ . \qquad (2.32)$$

The amplitude of vibration of the lever, $A_b(\omega, F_1)$, denoted now by the two parameters ω and F_1, will be given by

$$A_b(\omega, F_1)[k' - \omega^2 m + i\omega\gamma] = ak\exp(i\theta) \ , \qquad (2.33)$$

where $k' = k - F_1$. The physics in this case is somewhat more involved than that of the uniform force case, because the motion of the lever is now influenced by the force derivative. It is clear, for example, that for a negative force derivative, the lever, as it moves down in its excursion, will be assisted by the increasing interaction force and will move further down than the lever in the uniform interaction force case. As the lever moves up, the interaction force decreases so that the net restoring force acting on the lever increases and the lever will again move up, further than for the uniform force case. Consequently, the amplitude of vibration of the lever will increase. The opposite will happen if the interaction force derivative is positive, namely, when $k' > k$. When the value of k' varies significantly from the value of k, the motion of the lever will become anharmonic, unless we keep the amplitude of vibration small enough. This condition will be assumed in all the discussions that follow, so that we can still describe the motion of the lever by Eq. (2.7). The resonance frequency, namely, that frequency at which the amplitude is maximized, changes from ω_0 to ω'_0, where

$$\omega'_0 = \sqrt{k'/m} \ . \qquad (2.34)$$

We note that k', which is the effective spring constant, must remain positive. Using the definition of γ and ω'_0 in Eq. (2.33) gives for the amplitude of vibration in the presence of F_1

$$A_b(\omega, F_1) = \frac{a\omega_0^2}{[(\omega'^2_0 - \omega^2)^2 + \omega^2\omega^2_0/Q^2]^{1/2}} \ , \qquad (2.35)$$

which can also be written as

$$A_b(\omega, F_1) = a \frac{\omega_0^2}{\omega'^2_0} \frac{Q}{[Q^2(1 - \omega^2/\omega'^2_0)^2 + \omega^2\omega_0^2/\omega'^4_0]^{1/2}}, \tag{2.36}$$

or as

$$A_b(\omega, F_1) = a \frac{\omega_0^2}{\omega\omega'_0} \frac{Q}{[Q^2(\omega/\omega'_0 - \omega'_0/\omega)^2 + \omega_0^2/\omega'^2_0]^{1/2}}. \tag{2.37}$$

In terms of the bimorph-driven lever response function,

$$G_b(\omega, F_1) = \frac{\omega_0^2}{\omega'^2_0} \frac{Q}{[Q^2(1 - \omega^2/\omega'^2_0)^2 + \omega^2\omega_0^2/\omega'^4_0]^{1/2}}, \tag{2.38}$$

we get

$$A_b(\omega, F_1) = \mathrm{a}\, G_b(\omega, F_1) \tag{2.39}$$

for the amplitude,

$$\sin\theta_b = G_b(\omega, F_1) \frac{\omega}{Q\omega_0} \tag{2.40}$$

for the sine angle, and

$$\cos\theta_b = G_b(\omega, F_1) \frac{\omega'^2_0 - \omega^2}{\omega_0^2} \tag{2.41}$$

for the cosine angle. Note that Eqs. (35) through (41) involve both ω_0 and ω'_0 because of the presence of the force derivative F_1. Figure 2.4(a) shows $A(\omega, F_1)$ (full line) and $\partial A/\partial\omega$ (dashed line) as a function of ω for $F_1 = 0.5$ N/m, where we see that because $F_1 = 0.5$, the peak of the $A_b(\omega)$ and its derivative is shifted to a lower frequency, and the amplitude is increased by $1.42\times Q$ because of the presence of $F_1 = 0.5\ N/m$. Figure 2.4(b) shows $\sin\theta$ (full line) and $\cos\theta$ (dashed line) as a function of ω for $F_1 = 0.5$ N/m. The importance of showing both the sine and cosine functions is twofold. First, it helps to identify the proper angular quadrant of the phase angle θ. Second, we need to know θ because it is used by the phase-sensitive detector that measures the component of the amplitude relative to a reference signal. We see that the force derivative shifts the peak of $\sin\theta$ to ω'_0, at which frequency $\cos\theta$ changes sign. On resonance, where $\omega = \omega'_0$, we get

$$G_{b\,0} = Q\sqrt{\frac{k}{k'}}\ , \tag{2.42}$$

$$A_{b\,0} = aG_{b\,0}\ , \tag{2.43}$$

and

$$\theta_{b\,0} = \pi/2\ , \tag{2.44}$$

where $\sqrt{k/k'}$ is an amplification factor. Figures 2.4(c) and 2.4(d) show $A_b(\omega, F_1)$ (full line) and $\partial A_b/\partial F_1$ (dashed line), and $\sin\theta_b$ (full line) and $\cos\theta_b$ (dashed line), respectively, as a function of F_1 for $\omega = \omega_0$. As before, the derivative obtains its maximum value away from ω_0. As F_1 grows in either polarity, the system becomes less tuned and its amplitude of vibration decreases. The maximum value of the amplitude equals Q because for $F_1 = 0$ the resonance frequency $\omega'_0 = \omega_0$. Also, the sine function obtains its maximum value and the cosine function changes sign at $\omega = \omega_0$. We note that in an experiment, where we scan ω or F_1, we must fix the phase angle θ, since the result of the experiment will be the product of the amplitude with the sine or cosine functions for the chosen phase angle θ.

2.4.3. Approximations

We have analyzed the general response of the system as a function of frequency and force derivative as shown in Figs. 2.4(a) through 2.4(d). There are cases, however, where the force derivative is sufficiently small that we can obtain simple analytic expressions for the interdependence of $G_b(\omega)$, F_1, and ω_0. We will divide our discussion into two parts; in the first, $\omega = \omega_0$, and in the second, $\omega = \omega_m$, where ω_m is an optimized operating frequency.

2.4.4. Operation on Resonance

Here the lever is vibrated at its free-resonance frequency ω_0 in the presence of a force derivative that is sufficiently small that $\omega_0 \simeq \omega'_0$. The new resonance frequency, ω'_0, can be approximated using Eq. (2.34):

$$\omega'_0 \simeq \sqrt{\frac{k}{m}}\left(1 - \frac{F_1}{2k}\right), \tag{2.45}$$

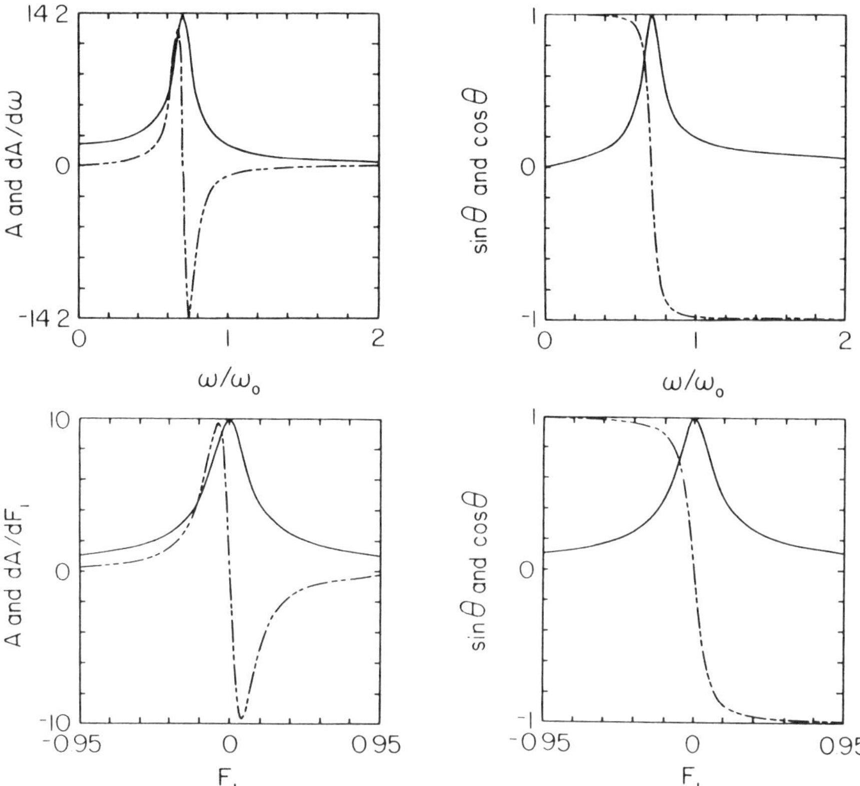

Fig. 2.4 Bimorph-driven lever: (a) $A_b(\omega)$ (full line) and $\partial A_b/\partial\omega$ (dashed line) as a function of ω/ω_0 for $F_1 = 0.5$ N/m. (b) $\sin\theta(\omega)$ (full line) and $\cos\theta(\omega)$ (dashed line) as a function of ω/ω_0 for $F_1 = 0.5$ N/m. (c) $A_b(F_1)$ (full line) and $\partial A_b/\partial F_1$ (dashed line) as a function of F_1 for $\omega = \omega_0$. (d) $\sin\theta(F_1)$ (full line) and $\cos\theta(F_1)$ (dashed line) as a function of F_1 for $\omega = \omega_0$.

so that

$$\frac{F_1}{2k} \simeq 1 - \frac{\omega'_0}{\omega_0} \tag{2.46}$$

and

$$\frac{\Delta\omega}{\omega_0} \simeq \frac{1}{2}\,\frac{F_1}{k}\,. \tag{2.47}$$

Note that a negative derivative reduces the effective spring constant while a positive one will increase it. From Eq. (2.27) and Fig. 2.2 we find that the lever, operating at ω_0 with $F_1 = 0$, has an amplitude

$$A_b(\omega_0, 0) = aQ \ . \tag{2.48}$$

Here the first and second terms in the parentheses denote the frequency of operation and the force derivative, respectively. The amplitude of vibration of the lever at its new resonance frequency, $\omega'_0(F_1)$, however, is given by

$$A_b(\omega'_0, F_1) = \mathrm{a} \, \frac{k}{k'} \, Q \ . \tag{2.49}$$

Using Eq. (2.48) and Eq. (2.49) in Eq. (2.37) gives for the scaled amplitude

$$r \simeq \frac{1}{[1 + Q^2 F_1{}^2/k^2]^{1/2}} \ , \tag{2.50}$$

where

$$r = \frac{A_b(\omega_0, F_1)}{A_b(\omega'_0, F_1)} \ . \tag{2.51}$$

Equation (2.50) gives the reduced amplitude of vibration of the lever operating at ω_0 in the presence of small F_1. By expanding r we get

$$\delta A_b(\omega_0) \simeq \frac{1}{2} \, A_b(\omega_0) \, \frac{Q^2}{k^2} \, F_1{}^2 \ . \tag{2.52}$$

As seen in Fig. 2.2 for $A_b(\omega_0, F_1)$, the effect of having a small F_1 gives rise to a negligible value for $\delta A_b(\omega_0)$.

Instead of vibrating the bimorph at the resonance frequency of the lever using an external oscillator, it is possible to use the lever as a crystal controlling an oscillator (Albrecht et al. 1990a). To that end the thermal vibrations of the lever are detected by a probe whose output is amplified, phase shifted, and fed back to the bimorph supporting the lever. The feedback-driven lever will then oscillate at a controlled constant amplitude A close to its natural frequency ω_0, with a full width at half maximum power given by (Heer 1972)

$$\Delta\omega = \omega_0 \, \frac{KT}{QkA^2} \ . \tag{2.53}$$

Here K is Boltzmann's constant, T is the temperature, and Q is the quality factor associated with the response function of the thermally vibrating lever. By enclosing the vibrating lever in vacuum (Israelachvili 1972), its natural Q can grow substantially ($Q \simeq 30{,}000$, Albrecht et al. 1990a). The oscillating lever will now have an effective Q given by (Albrecht et al. 1990b)

$$Q_{eff} = \frac{kA^2}{KT} Q \,. \tag{2.54}$$

Since $F_1 \simeq 2k\ \Delta\omega/\omega_0$, we find that by measuring the frequency of the feedback-driven lever we can measure the force derivative directly. The sensitivity of this FM method is approximately

$$F_1 \simeq \frac{2}{A} \sqrt{\frac{KTkB}{Q\omega_0}} \,. \tag{2.55}$$

The technique has the unique feature that the lever can respond immediately to changes in the frequency. This is in contrast to the case where the bimorph is driven by an external oscillator, where it takes a time on the order of $2\pi/\Delta\omega$ for the lever to change its amplitude of vibration when its resonance frequency changes.

2.4.5. Operation at the Optimum Frequency

As seen in Fig. 2.3, the maximum value of the derivatives of $A_b(\omega)$ in respect to ω is slightly off resonance. By equating to zero the second derivative of $A_b(\omega)$ in Eq. (2.39), this frequency, denoted by ω_m, is given by

$$\frac{|\omega_0 \pm \omega_{\pm m}|}{\omega_0} \simeq \frac{0.35}{Q} \,. \tag{2.56}$$

Inserting the value of ω_m in Eq. (2.23) gives for $A_b(\omega_{m,0})$

$$\frac{A_b(\omega_m, 0)}{A_b(\omega_0, 0)} \simeq 0.816 \,. \tag{2.57}$$

The variation of the amplitude in respect to ω and to δF_1 at this frequency yields (Martin et al. 1987)

$$\delta(A_b(\omega_m,0)) = \frac{4}{3\sqrt{3}} \frac{Q}{\omega_0} A_b(\omega_0,0)\,\delta\omega \tag{2.58}$$

and

$$\delta(A_b(\omega_m,F_1)) = \frac{2}{3\sqrt{3}} \frac{Q}{k} A_b(\omega_0,0)\,\delta F_1 \,. \tag{2.59}$$

Comparing Eq. (2.52) with Eq. (2.59) shows that operating at $\omega = \omega_m$ gives a first-order perturbation of $A_b(\omega_m)$ for δF_1, in contrast to the case where $\omega = \omega_0$ and the effect is of second order.

2.5. Sample-Driven Lever

Instead of vibrating the lever using a bimorph, we can set the lever in motion by a vibrating sample (Israelachvili 1972). Here the interaction between the tip mounted on the lever and the sample is by means of a force derivative in the direction of vibration of the lever. We find for the system, shown in Fig. 2.5, that

$$u = u_0 \,, \tag{2.60}$$

$$z = z_0 + \zeta \,, \tag{2.61}$$

and

$$g = g_0 + a\exp(i\omega t) \,. \tag{2.62}$$

For an amplitude of vibration that is sufficiently small, we can again expand the force around its average value and obtain

$$F(z-g) = F_0(z_0-g_0) + F_1(z_0-g_0)[\zeta - a\exp(i\omega t)] \,. \tag{2.63}$$

Note that because the force acts between the tip on the lever and the sample, its expansion is in terms of $\zeta - a\exp(i\omega t)$. The equation of motion in this case, which is given by

$$m\frac{\partial^2 z}{\partial t^2} + \gamma\frac{\partial z}{\partial t} + k(z-u) = F(z-g) \,, \tag{2.64}$$

is similar to the one for the bimorph-driven lever except that g, the position of the sample, is not zero. At equilibrium we find again that

$$F(z_0-g_0) = k(z_0 - u_0) \,. \tag{2.65}$$

Since the motion of the lever for a small amplitude can be considered harmonic, we still can write it using Eq. (2.17). As a result, Eq. (2.64) becomes

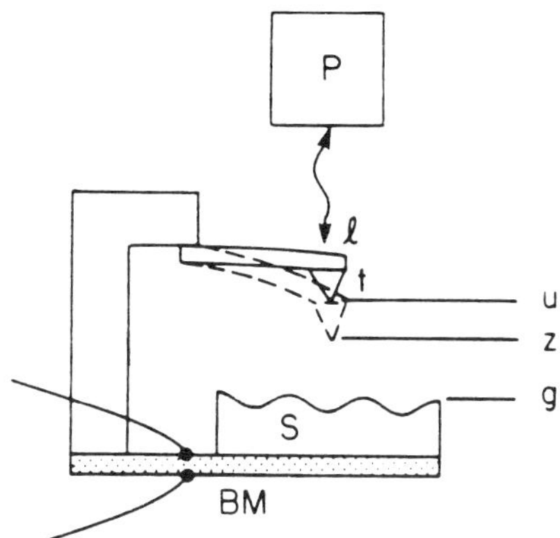

Fig. 2.5 The geometry of the sample-driven lever.

$$A_s(\omega, F_1)[k' - \omega^2 m + i\omega\gamma] = -aF_1 \exp(i\theta) \ , \tag{2.66}$$

which is similar to that for the bimorph-driven lever except that k is replaced by $-F_1$. Using the sample-driven response function $G_s(\omega, F_1)$,

$$G_s(\omega, F_1) = \frac{|F_1|}{k'} \frac{Q}{[Q^2(1 - \omega^2/\omega'^2)^2 + \omega^2\omega_0{}^2/\omega'_0{}^4]^{1/2}} \ , \tag{2.67}$$

we get

$$A_s(\omega, F_1) = \mathrm{a}\ G_s(\omega, F_1) \tag{2.68}$$

for the amplitude of vibration,

$$\sin\theta_s = -G_s(\omega, F_1) \frac{k'}{QF_1} \frac{\omega\omega_0}{\omega'_0{}^2} \tag{2.69}$$

for the sine function, and

$$\cos\theta_s = -G_s(\omega, F_1) \frac{k'}{F_1} \frac{\omega'_0{}^2 - \omega^2}{\omega'_0{}^2} \tag{2.70}$$

for the cosine function. Note that the amplitude $A_s(\omega, F_1)$ has an amplification factor that is now given by $|F_1|/(kk')^{1/2}$. Figures 2.6(a) and 2.6(b) show $A_s(\omega, F_1)$ (full line) and $\partial A_s/\partial\omega$ (dashed line), and $\sin\theta_s$ (full line) and $\cos\theta_s$ (dashed line), respectively, as a function of ω where $F_1 = 0.5$ N/m. Comparing Fig. 2.3(a) with Fig. 2.5(a) shows that the

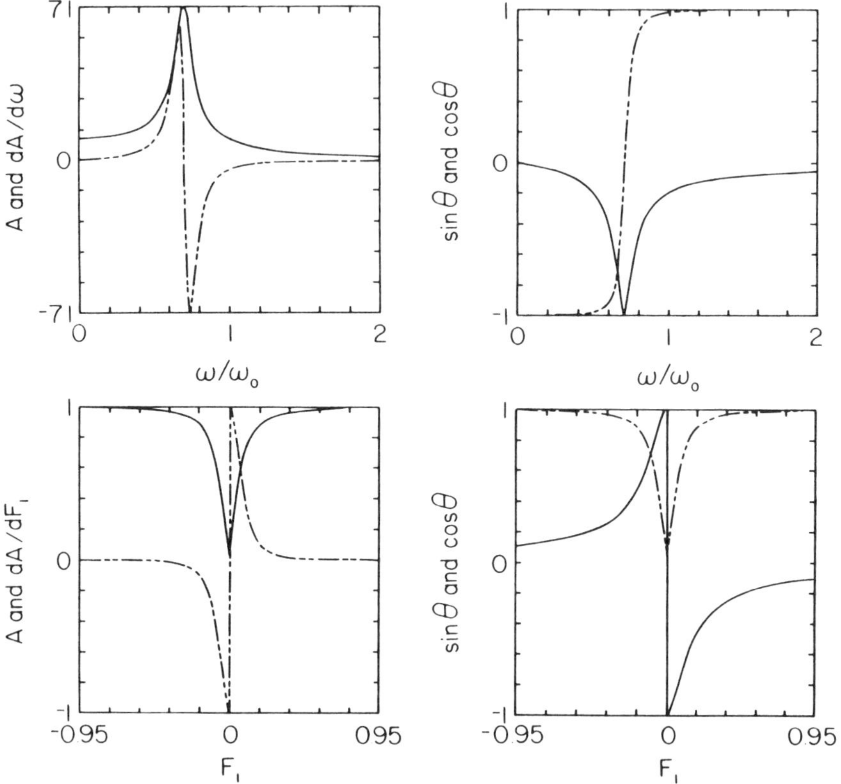

Fig. 2.6 Sample-driven lever: (a) $A_b(\omega)$ (full line) and $\partial A_b/\partial\omega$ (dashed line) as a function of ω/ω_0 for $F_1 = 0.5$ N/m. (b) $\sin\theta(\omega)$ (full line) and $\cos\theta(\omega)$ (dashed line) as a function of ω/ω_0 for $F_1 = 0.5$ N/m. (c) $A_b(F_1)$ (full line) and $\partial A_b/\partial F_1$ (dashed line) as a function of F_1 for $\omega = \omega_0$. (d) $\sin\theta(F_1)$ (full line) and $\cos\theta(F_1)$ (dashed line) as a function of F_1 for $\omega = \omega_0$.

maximum amplitude is now only 7.1 because of the amplification factor illustrated in Fig. 2.3(a). The sine and cosine functions are similar to those of Fig. 2.4 except that they are inverted. Figures 2.6(c) and 2.6(d) show $A_s(\omega, F_1)$ (full line) and $\partial A_s/\partial F_1$ (dashed line), and $\sin\theta_s$ (full line) and $\cos\theta_s$ (dashed line), respectively, as a function of F_1 where $\omega = \omega_0$. For $F_1 = 0$, the amplitude and its derivative are zero because in the presence of a uniform force, $F = F_0$, the sample cannot vibrate the lever. As F_1 grows in either polarity, the amplitude grows as well, and its derivative reaches its maximum for small values of F_1. The sine and cosine functions in this case are different from those relating to the bimorph-driven lever. If we operate at $\omega = \omega'_0(F_1)$, namely for each value of F_1 we tune the operating frequency to, we find that the amplitude and phase angle are

$$A_{s0} = aQ \frac{|F_1|}{\sqrt{kk'}} \tag{2.71}$$

and

$$\theta_{s0} = -\pi/2 \ , \tag{2.72}$$

respectively.

2.6. Tip-Driven Lever

This case, which is shown in Fig. 2.7, is similar to the sample-driven case except that here the probe measures the tip-sample distance rather than the position of the lever. Equations (2.60) through (2.72) still hold, but the parameter that we measure is

$$A_t(\omega, F_1)\exp(-i\theta_t) = a - A_s\exp(-i\theta_s) \ , \tag{2.73}$$

where A_s is given by Eq. (2.68). The tip-driven response function $G_t(\omega, F_1)$, defined by

$$A_t(\omega'_0, F_1) = aG_t(\omega'_0, F_1) \ , \tag{2.74}$$

is given by

$$G_t(\omega, F_1) = [1 - 2T_s(\omega, F_1)\cos\theta_s + T^2_s(\omega, F_1)]^{1/2} \ , \tag{2.75}$$

which can be simplified (Dürig and Zügar 1989) to

$$T_t(\omega, F_1) = \left[\frac{(\omega^2 - \omega_0{}^2)^2 + (\omega_0\omega/Q)^2}{(\omega^2 - \omega_0{}'^2)^2 + (\omega\omega_0/Q)^2}\right]^{1/2} . \tag{2.76}$$

For a general discussion, see for example Dürig et al. (1992). The sine and cosine functions can be written in terms of the sample-driven parameters by

$$\sin\theta_t = \frac{G_s(\omega, F_1)\ (\sin\theta_s)}{G_t(\omega, F_1)} \tag{2.77}$$

and

$$\cos\theta_t = \frac{[1 - G_s(\omega, F_1)\cos\theta_s]}{G_t(\omega, F_1)} . \tag{2.78}$$

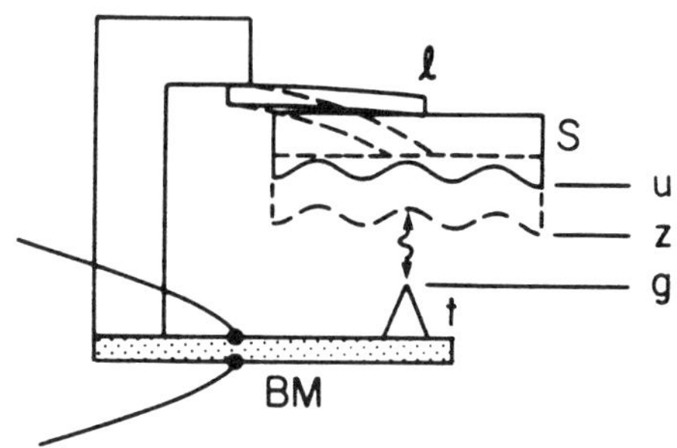

Fig. 2.7 The geometry of the tip-driven lever.

On resonance, where $\omega = \omega'_0$,

$$G_t(\omega'_0, F_1) = (1 + \sigma)^{1/2} , \tag{2.79}$$

$$\sin\theta_t = - \frac{\sigma^{1/2}}{(1 + \sigma)^{1/2}} , \tag{2.80}$$

and

$$\cos\theta_t = \frac{1}{(1 + \sigma)^{1/2}} , \tag{2.81}$$

where σ is given by

$$\sigma = Q^2 \frac{(k - k')^2}{kk'} . \tag{2.82}$$

At $\omega = \omega_0$,

$$G_t(\omega_0, F_1) = \frac{1}{(1 + \sigma k'/k)^{1/2}} \tag{2.83}$$

and

$$A_t(\omega_0, F_1) = aG_t(\omega_0, F_1) , \tag{2.84}$$

giving

$$\frac{G_t(\omega'_0, F_1)}{A_t(\omega_0, F_1)} = \left[\left(1 + \frac{\sigma k'}{k}\right)(1 + \sigma)\right]^{1/2} \tag{2.85}$$

for the ratio. Figures 2.8(a) and 2.8(b) show $A_t(\omega, F_1)$ (full line) and $\partial A_t/\partial\omega$ (dashed line) and $\sin\theta_t$ and $\cos\theta_t$ as a function of ω for $F_1 = 0.5$, and Figs. 2.8(c) and 2.8(d) show $A_t(\omega, F_1)$ (full line) and $\partial A_t/\partial F_1$ (dashed line) and $\sin\theta_t$ and $\cos\theta_t$ as a function of F_1 for $\omega = \omega_0$. The unique feature in the tip-driven case is that since the amplitude that we measure is the sum of two terms that are not necessarily in phase, the resultant amplitude as a function of ω obtains a minimum at ω_0 and a maximum of 7.3 at ω'_0, and the sine function has a "dimple" centered at ω'_0. Figure 2.8(c) shows that the amplitude obtains its maximum value at $F_1 = 0$, where the tip cannot vibrate the sample and the relative amplitude is therefore unity. For larger values of F_1, in either polarity, the amplitude would have increased except that the system is detuned by F_1. One can clearly increase the amplitude of vibrations by operating at $\omega = \omega'_0(F_1)$, as seen in Fig. 2.8(a).

2.7. Summary

We have calculated the response of a tip attached to a lever for three typical cases that probably cover most of the possible implementations of a scanning force microscope where the mechanical resonance of the lever is utilized. In later chapters we utilize the results we obtained for these three cases and discuss the merits of each one for measuring a particular force. We found the amplitude of vibration of the lever for the bimorph-driven case

$$A_b(\omega, F_1) = \mathrm{a}\,\frac{k}{k'}\,\frac{Q}{\sqrt{Q^2(1 - \omega^2/\omega_0'^2)^2 + \omega^2\omega_0^2/\omega_0'^4}},$$

the sample-driven case

$$A_s(\omega, F_1) = \mathrm{a}\,\frac{|F_1|}{k'}\,\frac{Q}{[Q^2(1 - \omega^2/\omega_0'^2)^2 + \omega^2\omega_0^2/\omega_0'^4]^{1/2}},$$

and the tip-driven case

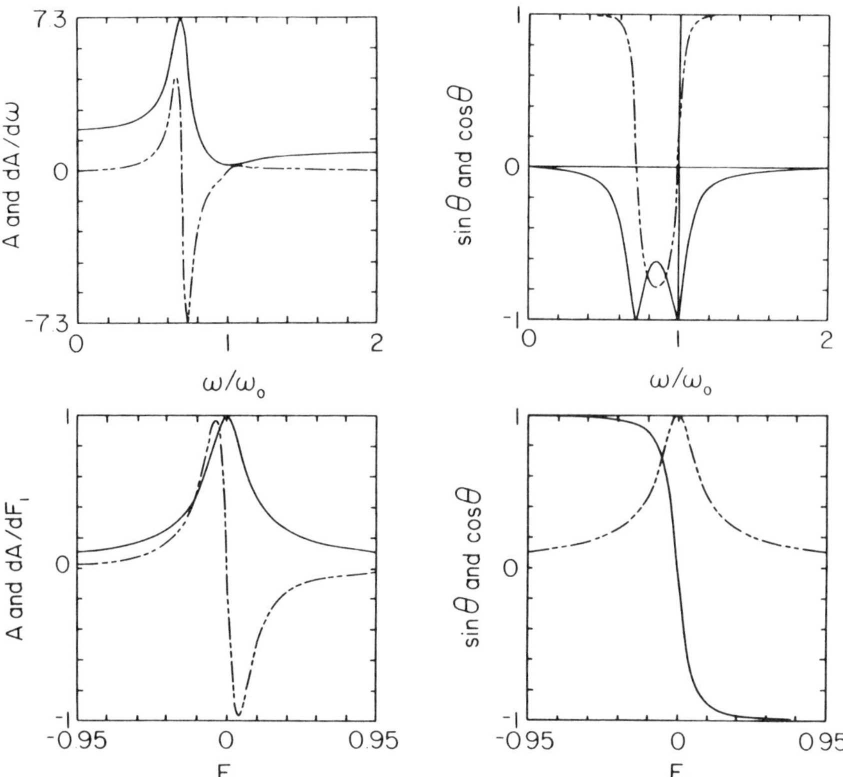

Fig. 2.8 Tip-driven lever: (a) $A_b(\omega)$ (full line) and $\partial A_b/\partial\omega$ (dashed line) as a function of ω/ω_0 for $F_1 = 0.5$ N/m. (b) $\sin\theta(\omega)$ (full line) and $\cos\theta(\omega)$ (dashed line) as a function of ω/ω_0 for $F_1 = 0.5$ N/m. (c) $A_b(F_1)$ (full line) and $\partial A_b/\partial F_1$ (dashed line) as a function of F_1 for $\omega = \omega_0$. (d) $\sin\theta(F_1)$ (full line) and $\cos\theta(F_1)$ (dashed line) as a function of F_1 for $\omega = \omega_0$.

$$A_t(\omega, F_1) = a\left[\frac{(\omega^2 - \omega_0{}^2)^2 + (\omega\omega_0/Q)^2}{(\omega^2 - \omega'_0{}^2)^2 + (\omega\omega_0/Q)^2}\right]^{1/2}$$

In using these expressions one must consider the free-lever resonance frequency ω_0, the shifted resonance frequency ω_0', the operating frequency ω, and the force derivative F_1.

3
Sources of Noise

3.1. Introduction

A typical scanning force microscope consists of a lever and its force-sensing tip, a bimorph, and a tip deflection probe with its amplifier, each of which contributes noise to the system. We present the primary sources of laser-diode noise (Petermann 1988 and Agrawal and Dutta 1986) including shot, intensity, and phase noise, which is converted into amplitude noise in an interferometric configuration. Next, we address the thermal noise of the lever and the limiting effect it has on the sensitivity of the measurements. We then discuss the noise of the bimorph supporting the lever and present experimental results where these noise contributions have been measured.

3.2. General Discussion of Noise

A time-dependent optical power, $P(t)$, can be written as a sum of two terms:

$$P(t) = P_0 + \delta P(t) \ , \tag{3.1}$$

where $P_0 = \langle P \rangle$ is the average power defined as

$$\langle P \rangle = \lim_{t' \to \infty} \frac{1}{t'} \int_{t_0}^{t_0 + t'} P(t)dt \ , \tag{3.2}$$

and $\delta P(t)$ is the noise term. The spectral density of the power, $W_p(\omega_m)$, where ω_m is the operating frequency, is defined by

$$W_p(\omega_m) = \langle |\Delta P(\omega_m)|^2 \rangle \ , \tag{3.3}$$

which can be obtained from $\delta P(t)$ in terms of a Fourier transform by

$$W_p(\omega_m) = \lim_{t' \to \infty} \frac{1}{t'} \left| \int_0^{t'} \delta P(t) \exp(-j\omega_m t) dt \right|^2 . \tag{3.4}$$

Here $\langle |\Delta P(\omega_m)|^2 \rangle$ is the mean square of the fluctuations, and the mean square of the noise, $\langle \delta P^2(t) \rangle$, is given by

$$\langle \delta P^2(t) \rangle = \int_{-\infty}^{\infty} W_p(\omega_m) df = \int_{-\infty}^{\infty} \langle |\Delta P(\omega_m)|^2 \rangle df . \tag{3.5}$$

The noise, passing through a filter that has a bandwidth of B and transmission of unity around the center frequency ω_m, is

$$\langle \delta P^2 \rangle = 2B \langle |\Delta P(\omega_m)|^2 \rangle = 2BW_p(\omega_m) . \tag{3.6}$$

A widely used definition of noise is given in terms of the relative intensity noise (RIN), defined by

$$\text{RIN} \equiv \frac{\langle \delta P^2 \rangle}{\langle P \rangle^2} . \tag{3.7}$$

Since an optical power incident on a photoconductive detector generates a current given by

$$i = \eta P , \tag{3.8}$$

where η is in mA/mW, we can write the RIN as

$$\text{RIN} \equiv \frac{\langle \delta i^2 \rangle}{\langle i \rangle^2} . \tag{3.9}$$

An advantage of using the RIN is that contributions from independent sources are additive. For example, we can define a signal-to-noise ratio (SNR) for an optical signal that is amplitude modulated by a sine wave. If the average optical power of the signal is $\langle P \rangle$ and that of its modulation is P', then the modulation index m is given by

$$m = \frac{P'}{\langle P \rangle} , \tag{3.10}$$

and the SNR becomes

$$\text{SNR} = \frac{1}{2} \frac{P'^2}{\langle \delta P^2 \rangle} = \frac{m^2}{2} \frac{\langle P \rangle^2}{\langle \delta P^2 \rangle} = \frac{m^2}{2RIN} . \tag{3.11}$$

3.3. Shot Noise

The RIN arising from statistics of photons incident on a photoconductive detector can be obtained from

$$\text{RIN} = \frac{2B\langle |\Delta P(\omega_m)|^2 \rangle}{\langle P^2 \rangle} = \frac{2BW_p(\omega_m)}{\langle P \rangle^2} . \tag{3.12}$$

From the definition of shot noise,

$$\langle |\Delta i(\omega_m)|^2 \rangle = e\langle i \rangle , \tag{3.13}$$

where e is the electronic charge, we get the shot-noise-limited RIN,

$$\text{RIN}_{shot} = \frac{2Be}{\langle i \rangle} = \frac{3.2\times10^{-16} mA}{\langle i \rangle} \frac{B}{Hz} . \tag{3.14}$$

The RIN of any experimental system will therefore obey

$$\text{RIN} > \text{RIN}_{shot} . \tag{3.15}$$

We find that a shot noise current

$$\langle \delta i^2 \rangle_{shot}{}^{1/2} = \sqrt{2e\eta P_0 B} \tag{3.16}$$

generates a voltage across a load resistor R given by

$$\langle \delta V^2 \rangle_{shot}{}^{1/2} = R\sqrt{2e\eta P_0 B} . \tag{3.17}$$

3.4. Resistor Johnson Noise

The agitation of electrons in the load resistor R in the first stage of a photodetector amplifier gives rise to *a* RIN, noise current, and noise voltage given by

$$\mathrm{RIN}_J = \frac{4KTB}{R\eta^2 P^2_0} , \tag{3.18}$$

$$\langle \delta i^2 \rangle_J{}^{1/2} = \sqrt{\frac{4KTB}{R}} , \tag{3.19}$$

and

$$\langle \delta V^2 \rangle_J{}^{1/2} = \sqrt{4KTBR} . \tag{3.20}$$

Here K is Boltzmann's constant and T is the absolute temperature.

3.5. Laser Intensity Noise

The dominant source of noise in laser diodes is spontaneous emission. The laser diode, which behaves like a thermal source, is actually an amplifier for this noise. Apart from the spontaneous-emission noise we have to take into account the mode-partition noise, mode-hopping noise, and $1/f$ noise. The theory of these contributions is lengthy and we will not deal with it here. Practically, we can use the specifications accompanying a laser diode that give the RIN measured in a given bandwidth around a center frequency. A typical value for the laser-intensity-related RIN, including the shot noise, is

$$\mathrm{RIN}_{int} = 10^{-13}\, B/[\mathrm{Hz}] . \tag{3.21}$$

The RIN, given in dB, is

$$\mathrm{RIN}_{int} = 10 \log\left[\frac{\langle \delta P^2 \rangle}{\langle P \rangle^2}\right] (\mathrm{dB}) , \tag{3.22}$$

and the SNR in dB is

$$\mathrm{SNR}_{int} = 10 \log\left[\frac{\langle P \rangle^2}{\langle \delta P^2 \rangle}\right] (\mathrm{dB}) . \tag{3.23}$$

The noise current and noise voltage are given, respectively, by

$$\langle \delta i^2 \rangle_{int}{}^{1/2} = \eta P_0 \sqrt{\mathrm{RIN}_{int}} \tag{3.24}$$

and

$$\langle \delta V^2 \rangle_{int}{}^{1/2} = \eta P_0 R \sqrt{\mathrm{RIN}_{int}} \;. \tag{3.25}$$

The intensity noise of a multimode laser is much smaller than that of a single-mode laser because of negative cross-correlation effects that exist between the different modes.

3.6. Laser Phase Noise

Fluctuations in the phase of an optical wave passing through an interferometer can be translated into intensity noise on a photodetector. To analyze this effect, we express the amplitude of the optical wave by

$$E(t) = \sqrt{S(t)}\, \exp[j\phi(t)] \;, \tag{3.26}$$

where

$$S(t) = |E(t)|^2 \tag{3.27}$$

is the scaled number of photons. Fluctuations in $S(t)$ represent the intensity noise, treated before, while fluctuations in ϕ represent the phase noise. As an example, we consider an interferometer having two arms with different optical path lengths. The waves traversing the two arms,

$$E_1 = A_1 E(t) \tag{3.28}$$

and

$$E_2 = A_2 E(t - \tau) \;, \tag{3.29}$$

recombine, and their interference transforms the phase fluctuations into intensity noise. Here A_1 and A_2 account for the different attenuation of the amplitude, and τ is the difference in their optical path length. If $A_1 \neq A_2$, the interference is not complete, and we can use the visibility of the interference fringes, V,

$$V = \frac{i_{max} - i_{min}}{i_{max} + i_{min}} , \tag{3.30}$$

where i_{max} and i_{min} denote the photocurrent at constructive and destructive interference, respectively. The interference of the two waves, for $A_1 = A_2 = A$, yields

$$\langle P(t) \rangle = 2|A|^2 \langle |E(t)|^2 \rangle [1 + \mathrm{Re}\, \gamma(\tau)] . \tag{3.31}$$

Here $\gamma(\tau)$ is the complex degree of coherence

$$\gamma(\tau) = \frac{\langle E(t) E^*(t - \tau) \rangle}{\langle |E(t)|^2 \rangle} , \tag{3.32}$$

which for a monochromatic light is

$$\gamma(\tau) = \exp(j2\pi\nu\tau) . \tag{3.33}$$

The coherence time τ_c can be written as

$$\tau_c = \frac{2|\tau|}{\langle \Delta\phi^2 \rangle} , \tag{3.34}$$

or as

$$\tau_c = \int_{-\infty}^{\infty} |\gamma(\tau)|^2 d\tau , \tag{3.35}$$

which gives $\langle \Delta\phi^2 \rangle^{1/2} = \sqrt{2}$ radians for $\tau = \tau_c$. The coherence length can be written as

$$L_c = \frac{c}{\Delta\nu} = c\tau_c , \tag{3.36}$$

where $\Delta\nu$ is the linewidth of the laser. The average optical power for the interference of a monochromatic beam in a two-arm interferometer is

$$\langle P_T(t) \rangle = 2|A|^2 \langle |E(t)|^2 \rangle [1 + \cos(2\pi\nu\tau)] , \tag{3.37}$$

and the phase noise is converted by the interferometer into intensity noise given by

$$P_T(t) = P_{T0}[1 + \cos(\phi_0 + \Delta\phi(t,\tau))] \; . \tag{3.38}$$

For $\tau << \tau_c$, we find that

$$P_T(t) = P_{T0}[1 + \cos(\phi_0) - \Delta\phi(t,\tau)\sin\phi_0] \; . \tag{3.39}$$

Choosing a phase shift $\phi_0 = -\pi/2$ yields a maximum phase-intensity noise given by

$$P_T(t) = P_{T0}[1 + \Delta\phi(t,\tau)] \; . \tag{3.40}$$

The spectral density of the phase fluctuations, as measured by a spectrum analyzer, $W_{\Delta\phi}(\omega_m)$, is given in terms of the spectral density of the frequency noise, $W_{\phi}(\omega_m)$, by

$$W_{\Delta\phi}(\omega_m) = W_{\phi}(\omega_m)\,\frac{\sin^2(\omega_m\tau/2)}{(\omega_m/2)^2} \; . \tag{3.41}$$

From the definition for the RIN,

$$\mathrm{RIN}_{phase} = \frac{\langle|\Delta P_T(\omega_m)|^2\rangle 2B}{P^2{}_{T0}} = 2BW_{\Delta\phi} \; , \tag{3.42}$$

and using the relationship

$$W_{\Delta\phi} = \tau^2 W_{\phi} = 2\pi\tau^2\Delta\nu \; , \tag{3.43}$$

where $\Delta\nu$ is the linewidth of the laser, we get

$$\mathrm{RIN}_{phase} = 4\pi\Delta\nu B\tau^2 \; . \tag{3.44}$$

Consequently, the current and voltage phase noise become, respectively,

$$\langle\delta i^2\rangle_{phase}{}^{1/2} = \eta P_0\sqrt{4\pi\Delta\nu B}\;\tau \tag{3.45}$$

and

$$\langle\delta V^2\rangle_{phase}{}^{1/2} = \eta P_0 R\sqrt{4\pi\Delta\nu B}\;\tau \; . \tag{3.46}$$

Note that the RIN is proportional to the laser linewidth and that the noise current and noise voltage are proportional to $\Delta\nu^{1/2}$. A calculation of the linewidth of a single-mode laser with length ℓ and cavity absorption α (Miles et al. 1980) gives

$$\Delta\nu = 4\pi N(\Delta\nu_{cav})^2 \frac{h\nu_0}{P_0} . \tag{3.47}$$

Here $1 < N < 10$ is proportional to the population inversion, $\Delta\nu_{cav}$ is the passive cavity linewidth given by

$$\Delta\nu_{cav} = \frac{c}{2\pi n'\ell} \frac{1 - Re^{-\alpha\ell}}{R^{1/2}e^{-\alpha\ell/2}} , \tag{3.48}$$

and R is the facet reflectivity of the laser. The effective refractive index of the laser waveguide can be calculated from

$$n' = n\left[1 - \frac{\lambda_0}{n}\frac{\partial n}{\partial\lambda}\right] , \tag{3.49}$$

where n is the refractive index of the passive waveguide at wavelength λ_0.

3.7. Thermally Induced Lever Noise

The lever of a scanning force microscope acts as an isolated one-dimensional harmonic oscillator whose energy is

$$W = \frac{1}{2} m \left(\frac{\partial z}{\partial t}\right)^2 + \frac{1}{2} m\omega_0^2 z^2 . \tag{3.50}$$

Here each of the two degrees of freedom contributes $1/2\ KT$ to the thermal energy of the lever, giving a total thermal energy of KT. The spectral density of the thermal noise of this free-running lever (Heer 1972) is

$$W_p(\omega) = \frac{2\gamma KT}{m^2(\omega_0^2 - \omega^2)^2 + \gamma^2\omega^2} , \tag{3.51}$$

where $\gamma = m\omega_0/Q$ and $\omega_0^2 = k/m$. Equation (3.51), rewritten as

$$2B\, W_p(\omega) = \frac{4KTBQ}{k\omega_0} \frac{1}{Q^2(1 - \omega^2/\omega_0^2)^2 + \omega^2/\omega_0^2} , \tag{3.52}$$

can be compared to the equation that describes the amplitude of vibration of the bimorph-driven lever, which, for $k' \simeq k$, is

$$A^2_b(\omega) = \frac{a^2 Q^2}{Q^2(1 - \omega^2/\omega_0{}^2)^2 + \omega^2/\omega_0{}^2} . \tag{3.53}$$

Dividing Eq. (3.52) by Eq. (3.53) gives

$$\frac{A^2_b(\omega)}{2BW_p(\omega)} = \frac{a^2 Q k \omega_0}{4KTB} . \tag{3.54}$$

The RMS noise of the lever, obtained from the thermal noise spectral density, is

$$\langle \delta z^2 \rangle^{1/2} = \sqrt{2BW_p(\omega)} , \tag{3.55}$$

where B is the bandwidth of the spectrum analyzer, gives the signal-to-noise ratio

$$\mathrm{SNR} = \frac{\delta A_b(\omega)}{\langle \delta z^2 \rangle^{1/2}} , \tag{3.56}$$

which can be written as

$$\mathrm{SNR} = \mathrm{a} \sqrt{\frac{Q k \omega_0}{4KTB}} . \tag{3.57}$$

Note that Eq. (3.57) is a general result that is independent of frequency, provided the force driving the lever is constant. The RMS noise of the lever can be obtained by using Eq. (3.55), yielding

$$\langle \delta z^2 \rangle^{1/2} = \left[\frac{4kTB}{k\omega_0} \frac{Q}{Q^2(1 - \omega^2/\omega_0{}^2)^2 + \omega^2/\omega_0{}^2} \right]^{1/2} \tag{3.58}$$

In the presence of force derivatives, Eq. (3.58), for example, should include $G(\omega, F_1)$ instead of $G(\omega)$. On and below resonance we get therefore, respectively (Dürig et al. 1986, McClelland et al. 1987, Martin et al. 1987),

$$\langle \delta z^2(\omega = \omega_0) \rangle^{1/2} = \sqrt{\frac{4KTBQ}{k\omega_0}} \tag{3.59}$$

and

$$\langle \delta z^2(\omega < \omega_0) \rangle^{1/2} = \sqrt{\frac{4KTB}{Qk\omega_0}} \; . \tag{3.60}$$

In contrast to Section 3.6, where we dealt with fluctuations in the optical phase, here we have fluctuation in the optical path length of one arm of an interferometer. Equations (3.26) to (3.33) and Eqs. (3.38) to (3.40) also apply to our case, and from Eq. (3.40) we find

$$\mathrm{RIN}_{lever} = \langle \Delta\phi(t,\tau)^2 \rangle \; . \tag{3.61}$$

Setting

$$\langle \Delta\phi^2 \rangle^{1/2} = \frac{4\pi}{\lambda} \langle \delta \ell^2 \rangle^{1/2} \; , \tag{3.62}$$

and using Eq. (3.59), we get for operation on and off the lever resonance,

$$\mathrm{RIN}_{lever}(\omega = \omega_0) = \frac{64\pi^2}{\lambda^2} \frac{KTBQ}{k\omega_0} \tag{3.63}$$

and

$$\mathrm{RIN}_{lever}(\omega < \omega_0) = \frac{64\pi^2}{\lambda^2} \frac{KTB}{Qk\omega_0} \; . \tag{3.64}$$

The noise current and noise voltage for operation on and off resonance are given by

$$\langle \delta i^2(\omega_0) \rangle_{lever}{}^{1/2} = \frac{4\pi}{\lambda} \eta P_0 \sqrt{\frac{4KTBQ}{\omega_0 k}} \; , \tag{3.65}$$

$$\langle \delta i^2(\omega < \omega_0) \rangle_{lever}{}^{1/2} = \frac{4\pi}{\lambda} \eta P_0 \sqrt{\frac{4KTB}{\omega_0 Qk}} \; , \tag{3.66}$$

$$\langle \delta V^2(\omega_0) \rangle_{lever}{}^{1/2} = \frac{4\pi}{\lambda} \eta P_0 R \sqrt{\frac{4KTBQ}{\omega_0 k}} \,, \tag{3.67}$$

and

$$\langle \delta V^2(\omega < \omega_0) \rangle_{lever}{}^{1/2} = \frac{4\pi}{\lambda} \eta P_0 R \sqrt{\frac{4KTB}{\omega_0 Q k}} \,. \tag{3.68}$$

3.8. Bimorph Noise

The amplitude noise of a thin bimorph, using Eq. (2.1), can be written as

$$\langle \delta z^2 \rangle^{1/2}{}_{bm} = 3\beta \mathrm{d}_{31} \, [\ell_c / t]^2 \, \langle \delta V^2 \rangle^{1/2} \,, \tag{3.69}$$

where we assumed that $\delta/t << 1$, and used $\delta V = t_c \delta E$ for the driving voltage. A noise, δV, in the voltage driving the bimorph supporting the lever will be translated into an amplitude noise of the lever

$$\langle \delta z^2 \rangle_{bm}{}^{1/2} = 3 \, G(\omega, F_1) \, \beta \mathrm{d}_{31} \, [\ell_c / t]^2 \, \langle \delta V^2 \rangle^2 \,, \tag{3.70}$$

where $G(\omega, F_1)$ is the response function. At the resonance frequency, for example, $G(\omega, F_1) = Q$, while below resonance, $G(\omega, F_1) = 1/Q$.

3.9. Lever Noise-Limited SNR

We will now calculate the signal-to-noise ratio of a system employing a vibrating lever in the presence of its thermal vibrations for $F_1 << k$. We treat two cases: in (a) the lever is vibrated by a bimorph at the steepest slope of its mechanical response; and in (b) the sample, which is vibrated by a bimorph, vibrates the lever by means of the force derivative. In case (a), the application of the force derivative changes the amplitude of vibration by changing the resonance frequency of the lever. Dividing the change in the amplitude of vibration of the lever by the noise amplitude gives the lever noise-limited minimum force derivative (Martin et al. 1987)

$$F_{1min} = \frac{1}{A_0} \sqrt{\frac{27KTBk}{\omega_0 Q}} \,. \tag{3.71}$$

For case (b), we use Eq. (2.71) and divide the amplitude of the vibrating lever at its resonance frequency in the presence of a small F_1 by its noise amplitude, Eq. (3.58), and get for the SNR

$$\mathrm{SNR} = a\, F_1 \sqrt{\frac{\omega_0 Q}{4KTk'B}} \,. \tag{3.72}$$

For a SNR = 1, we find that the lever noise-limited minimum force derivative is

$$F_1 = \frac{1}{a} \sqrt{\frac{4KTk'B}{\omega_0 Q}} \,. \tag{3.73}$$

We see that for both cases the sensitivity of the system to a small F_1 improves by using a small k' and a large value for a and for Q.

Thermal noise is equally divided by amplitude and phase-noise contributions (Robins 1982). We can obtain the force derivative directly by measuring the frequency of the thermally driven lever. The minimum detectable force derivative, limited by the phase noise of the lever, is similar to that given by Eq. (3.72) (Albrecht, private communication).

3.10. Experimental Characterization of Noise

3.10.1. Noise in Scanning Tunneling Microscopy

The spectrum of noise of the tunneling current in a well-isolated STM under ultrahigh vacuum (UHV) conditions (Dürig et al. 1986) was used to measure the forces acting between the tip and the sample. The noise, for frequencies above 100 Hz, was found to be less than 1 pm/Hz$^{1/2}$ with peaks that resulted from the thermal vibration of the tunneling tip. The noise of a STM has since been extensively studied, and methods aimed at reducing it have been explored (Pohl 1986, Park and Quate 1987, Abraham et al. 1988c,d, Kuk and Silverman 1989, and Gimzewski et al. 1987). The noise generated by the environment, which consists of mechanical vibrations coupled into the STM, can be minimized using springs of various sorts that have a critical damping coefficient at low frequencies. The theory presented in Chapter 2 can be used to determine the effect of a damped spring in an isolation system on the transfer of mechanical vibrations from the environment to the STM. Use of single and double springs and magnetic damping for UHV operation, and bungy cords and "slime" material for operation in air, have practically eliminated the effect of the environment on the operation of the STM. Intrinsic noise of the STM, after elimination of externally coupled noise, has been predominantly the $1/f$ noise. For example, Abraham et al. (1988c) find that the current noise follows a $1/f^{0.94}$ curve with a noise of 100 $fA/Hz^{1/2}$ at 100 kHz, which could be reduced by using a dithering method. Möller et al. (1990) find that the noise follows a $1/f^{1.04}$ curve with a noise of 18 $fA/Hz^{1/2}$ at around 1 kHz .

3.10.2. Laser-Diode Noise

Optical methods have proven to be a powerful technique for probing the deflection of levers in scanning force microscopes, and laser diodes have recently replaced HeNe lasers for some applications. A comprehensive review of laser-diode modulation and noise has been presented by Petermann (1988), in which the theories of amplitude and phase noise, relevant to interferometric systems, have been discussed. A comparison between noise spectra of two Tropel HeNe lasers and Hitachi HLP 1400 laser diodes in a bandwidth of 30 Hz was investigated by Dandridge et al. (1980a). The measurements, which were carried out as a function of frequency, show that HeNe lasers had a constant noise up to a frequency of 100 kHz, above which the noise dropped. Laser-diode noise, however, is larger than that of stabilized HeNe lasers at frequencies below approximately 100 kHz, above which their noise becomes comparable. They also observed, using a stabilized HeNe laser, that the amplitude noise, which was flat over an optical power ranging from 1 μW to 1 mW, was equal to the shot noise at approximately 10 μW. For laser diodes, it was found that the relative noise was at a maximum around threshold and decreased as the power of the laser was increased. Miles et al. (1980) report an investigation of a Hitachi HLP laser having the following parameters pertaining to Eq. (3.48): $R = 0.32$, $\ell = 300\ \mu$m, $P_0 = 10$ mW, $n' = 4.2$, and $\alpha = 7\ \text{cm}^{-1}$, giving a bandwidth of $0.3 < \Delta\nu < 3$ MHz. The coherence length for this particular laser was found, therefore, to be in the range of 10 to 100 m. Multimode lasers, however, have a much smaller coherence length. The variation of noise with frequency at different driving current levels and at a 30-Hz bandwidth has been investigated by Dandridge et al. (1980b). They find that for several levels of laser-driving current, away from the threshold current where the RIN is maximum, the noise decreased linearly as the operating frequency and the driving current increased. Dandridge et al. (1981) report interferometric measurements of phase noise performed in quadrature using a single-mode laser diode for optical path-length difference (OPD) ranging from 100 μm to 1 m for modulation frequencies of 50 Hz, 500 Hz, and 2 kHz. The results show that the noise is linear in the OPD, in agreement with a model that describes the noise as arising from random fluctuations of the wavelength of the laser. The interferometer, for a zero OPD and a bandwidth of 1 Hz around 1 kHz, could resolve a phase shift of 10^{-7} rad. Similar results have been obtained by Rugar et al. (1989), who observed that the noise increased tenfold when the OPD changed from 0 to 14 μm.

3.10.3. Noise in Scanning Force Microscopy

Tables 3.1 and 3.2 present experimental results using HeNe lasers and laser diodes, respectively, in a scanning force microscope. Schönenberger and Alvarado (1989) measured a noise of 0.04 nm in the range 0.01 to 1 Hz and a noise of 0.001 nm in the range 1 to 20 kHz, and Rugar et al. (1989) measured a noise of 10^{-2} nm in the range 0 to 1 kHz. The effect of squeezing of lever thermal noise has been explored by Rugar and Gütter (1990, 1991), where thermal noise was reduced at the expense of the other quadrature by almost a factor of 2.

Table 3.1 Noise level using a HeNe laser.

Reference	Noise nm/Hz$^{1/2}$	Frequency
Meyer et al. (1988)	4×10^{-5} nm/Hz$^{1/2}$	10 kHz
Rugar et al. (1988)	1.7×10^{-5} nm/Hz$^{1/2}$	above 2 kHz
Schönenberger et al. (1990)	6×10^{-6} nm/Hz$^{1/2}$	1 kHz

Table 3.2 Noise level using a laser diode.

Reference	Noise nm/Hz$^{1/2}$	Frequency
Dandridge et al. (1980a)	3×10^{-4} nm/Hz$^{1/2}$	100 Hz
	9×10^{-5} nm/Hz$^{1/2}$	1 kHz
Sarid et al. (1990)	3×10^{-4} nm/Hz$^{1/2}$	1 kHz
Rugar et al. (1989)	5.5×10^{-5} nm/Hz$^{1/2}$	above 1 kHz

3.11. Summary

The main results obtained in this chapter are the contributions of the shot noise

$$\langle \delta i^2 \rangle_{shot}^{1/2} = \sqrt{2e\eta P_0 B} \, ,$$

the Johnson noise of the load resistor

$$\langle \delta i^2 \rangle_J^{1/2} = \sqrt{\frac{4KTB}{R}} \, ,$$

the relative intensity noise of the laser

$$\langle \delta i^2 \rangle_{int}^{1/2} = \eta P_0 \sqrt{\mathrm{RIN}_{int}} \, ,$$

the phase noise of the laser

$$\langle \delta i^2 \rangle_{phase}^{1/2} = \eta P_0 \sqrt{4\pi \Delta\nu B \tau} \, ,$$

the thermal noise of the lever on resonance

$$\langle \delta i^2(\omega_0) \rangle_{lever}^{1/2} = \frac{4\pi}{\lambda} \eta P_0 \sqrt{\frac{4KTBQ}{k\omega_0}} \, ,$$

and the thermal noise of the lever off resonance

$$\langle \delta i^2(\omega < \omega_0) \rangle_{lever}^{1/2} {}^{1/2} = \frac{4\pi}{\lambda} \eta P_0 \sqrt{\frac{4KTB}{Qk\omega_0}} \, .$$

4
Tunneling Detection System

4.1. Introduction

In the following chapters we describe seven different types of atomic force microscopes that map forces across the surface of a sample. The development of these systems is an exercise in ingenuity, where many possible configurations have been tried. Common to each system is the use of a delicate sensor that measures the minute force-induced deflections of a flexible lever supporting the force-sensing tip. By raster scanning the tip across a surface, we get a two-dimensional mapping of the forces acting between the tip and the sample.

4.2. Theory

This chapter describes several systems that utilize tunneling of electrons to measure the deflection of the lever. The tunneling occurs between an auxiliary conducting tip and the lever separated by several angstroms. The application of a bias voltage between the tunneling tip and the lever produces a tunneling current through the air gap separating the two. Minute deflection of the lever supporting the force-sensing tip will vary the gap between the tunneling tip and lever and produces a change in the tunneling current. This current is eventually used to track and control the position of the lever, producing an image of the force distribution across the surface of the sample. A rough approximation to the tunneling current density j (Binnig and Rohrer 1986) gives

$$j = \frac{e^2}{\hbar}\,\frac{\kappa_0}{4\pi^2 z}\,V \exp(-2\kappa_0 z) \,. \tag{4.1}$$

Here z is the effective tunneling distance in Å, V is the bias voltage, and $e^2/\hbar = 2.44\times10^{-4}\ \Omega^{-1}$. Also, κ_0 is the inverse decay length of the wave function density outside the surface given by

$$2\kappa_0(\text{Å}) = 1.025\,\sqrt{\phi(eV)}\,, \tag{4.2}$$

and $\phi = (\phi_1 + \phi_2)/2$ is the effective barrier height. As an example, the tunneling current between a clean iridium tip and polycrystalline Ag under UHV conditions with V = -20 mV gives approximately (Gimzewski et al. 1987, Gimzewski and Möller 1987)

$$\log i(\text{nA}) = -0.66\, z(\text{Å}) \,. \tag{4.3}$$

By modulating the tunneling current, it is possible to measure displacements as small as 10^{-4} Å, corresponding to a force of 10^{-16} N, and by vibrating the lever, the sensitivity to force can be increased (Binnig and Rohrer 1986). The first atomic force microscope, reported by Binnig and Rohrer (1986), was an implementation of their scanning tunneling microscope. Their system, which operated in the ac mode, was capable of measuring forces on surfaces of insulators using a probe that did not damage the surface. The significance of this system derives from its being the original design and its capability to image the attractive van der Waals forces across dielectric surfaces with a high, though not atomic, resolution. Subsequently, Binnig et al. (1987) refined their system to deliver atomically resolved images in the repulsive mode. They mapped highly oriented pyrolytic graphite (HOPG) surfaces with an insulating stylus, achieving atomic resolution. This system was similar to their first one, and the improvement in resolution is attributed to the micromachined silicon-oxide lever, the edge of which served as the force-sensing tip. The AFM registered a modulation on the HOPG surface of 0.1 Å, or 10% of its value, while that of a typical STM was 1 Å, or 90%. Kirk et al. (1988) demonstrated the first atomic force microscope operating at liquid-He temperatures, obtaining atomic resolution of 2H-MoS_2. Marti et al. (1987) were the first to demonstrate atomic resolution of HOPG covered by oil, and Dürig et al. (1986) were the first to measure atomic forces acting on the tunneling tip from the spectrum of the tunneling current. Ultrahigh vacuum operation using a tunneling detection system has recently been reported by Bürgler et al. (1990), and a low-temperature system by Kirk et al. (1988), Probst et al. (1990), and Giessibl et al. (1990).

However, several difficulties have been encountered when using a tunneling tip to monitor the deflection of the force-sensing lever. For example, the tunneling process is sensitive to the particular atom or atoms between which tunneling takes place, and the presence of contaminants on the surface of the lever or on the apex of the tunneling tip alters the tunneling current dramatically. Such a strong dependence of the tunneling current on the atomic type of the tunneling-sensing tip and force-sensing tip can be detrimental to the accurate measurement of the deflection of the force-sensing lever. Even under ideal conditions, namely ultrahigh vacuum and clean surfaces, the tunneling tip will always drift across the lever and sense its atomic structure. As a result, the topographic image of the lever will be superimposed on the topographic image of the force. Another difficulty arises because of the proximity of the tunneling tip and the force-sensing lever between which atomic forces exist. Consequently, the tunneling current will be a measure of the force acting between the tunneling tip and force-sensing lever as well as between the force-sensing lever and sample. These difficulties led to the development of alternative methods to monitor the deflection of the lever. These methods use nonconducting, remote-sensing, and distance-averaging methods that eliminate many of the drawbacks of the tunneling method. In spite of this, the tunneling method has been very successful in producing atomically resolved images of conductors, semiconductors, and insulators.

We now turn the discussion to the four key components comprising a typical scanning force microscope using electron tunneling. The components, which are mounted on piezoelectric positioning elements, consist of a tunneling tip (T), a force-sensing tip (t) and its supporting lever (ℓ), and a sample (S). We classify the implementations of these force microscopes into five slightly different arrangements of the force-sensing and current-sensing elements. The arrangements consist of perpendicular, crossed, parallel, serial, and single-lever systems. For each class of instruments we present a brief discussion consisting of lever and tip specifications and performance highlights. Within a class, we group together the results according to the researchers that developed them.

4.3. Perpendicular Arrangement

Figure 4.1(a) shows a schematic of the original system where the lever supporting the force-sensing tip is placed perpendicular to the tunneling tip. Tables 4.1 through 4.13 illustrate lever and tip specifications and performance highlights of various researchers.

Table 4.1 Atomic force microscopy (Binnig and Rohrer 1986).

Lever and tip specifications	Performance highlights
lever material: gold foil	amplitude of vibration: 0.1 - 10 Å
lever shape: rectangular	smallest force: 10^{-9} N
lever length: 0.8 mm	smallest lever deflection: 1 Å
lever width: 250 μm	spatial resolution: 30 Å
lever thickness: 25 μm	environment: air
lever spring constant: 100 N/m	mode: ac
force-sensing tip: shredded diamond	force: atomic
tunneling tip: tungsten	
sample: ceramic Al_2O_3	

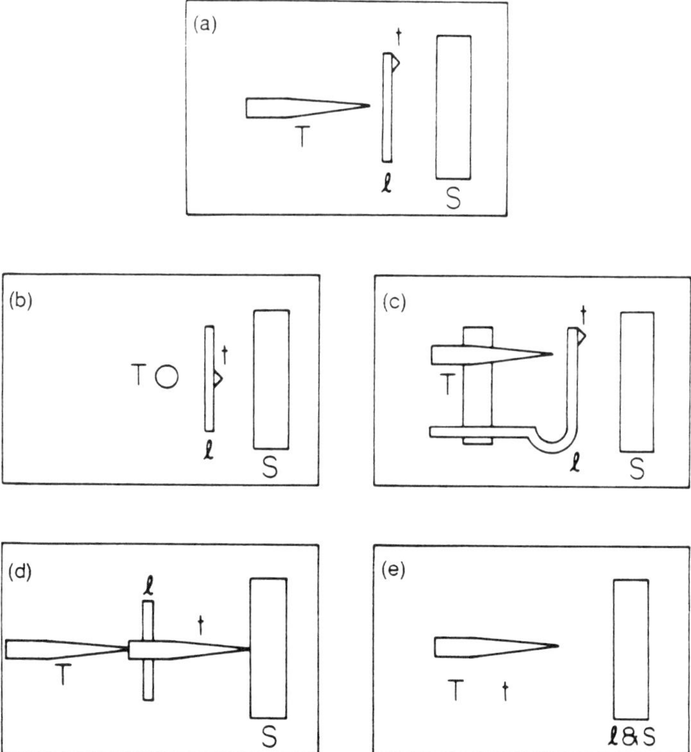

Fig. 4.1 The five geometries of arranging the tunneling tip (T), lever (ℓ) supporting the force-sensing tip (t), and sample (S). The geometries refer to the position of the lever (ℓ) supporting the force-sensing tip (t) relative to the tunneling tip (T) where ℓ and T are (a) perpendicular, (b) crossed, (c) parallel, (d) serial, and (e) where the force-sensing tip serves also as the tunneling tip.

Table 4.2 Atomic force microscopy (Binnig et al. 1987).

Lever and tip specifications	Performance highlights
lever material: SiO_2	typical force: 10^{-8} N
lever coating: 50 Å Cr +300 Å Au	thermal excitation: 3.5 Å
lever shape: rectangular	spatial resolution: atomic
lever length: 240 (180) μm	environment: air
lever width: 20 μm	mode: ac
lever thickness: 1.5 μm	force: atomic repulsive
lever spring constant: 0.07 (0.16) N/m	
lever resonance frequency: 21 (38) kHz	
force-sensing tip: SiO_2	
tunneling tip: tungsten	
sample: HOPG	

Table 4.3 Atomic force microscopy (Albrecht and Quate 1987, Albrecht and Quate 1988, and Kirk et al. 1988).

Lever and tip specifications	Performance highlights
lever material: SiO_2	typical force: 2×10^{-7} N
lever shape: V	typical lever deflection: 1000 Å
lever coating: 400 Å Au	spatial resolution: atomic
lever length: 100 μm	environment: UHV, T, air-water-oil
lever width: 22 μm	environment: air and He temperature
lever thickness: 1.5 μm	force: atomic repulsive
lever spring constant: 2 N/m	mode: ac
lever resonance frequency: 100 kHz	
force-sensing tip: SiO_2	
sample: HOPBN, HOPG, MoS_2, PODA, and PMMA	

Table 4.4 Atomic force microscopy (Heinzelmann et al. 1987).

Lever and tip specifications	Performance highlights
lever material: metallic glass	spatial resolution: 30 Å
lever shape: rectangular	environment: air
lever length: 700 μm	mode: dc
lever width: 50 μm	force: atomic repulsive
lever thickness: 15 μm	
lever resonance frequency: 20 kHz	
force-sensing tip: metallic glass	
tunneling tip: tungsten	
sample: quartz, HOPG, oxidized Si, PbS	

Table 4.5 Atomic force microscopy (Heinzelmann et al. 1987).

Lever and tip specifications	Performance highlights
lever material: metal foil	typical force: 10^{-7} N
lever spring constant: 0.1 - 10 N/m	spatial resolution: 2.5 Å
lever resonance frequency: 20 kHz	environment: air
force-sensing tip: metal foil	mode: dc
sample: HOPG, quartz, LiF, mineral glass	force: atomic attractive and repulsive

Table 4.6 Atomic force microscopy (Meyer et al. 1988 and Heinzelmann et al. 1989a).

Lever and tip specifications	Performance highlights
lever material: SiO_2	typical force: 10^{-8} N
lever shape: rectangular and triangular	spatial resolution: atomic
lever spring constant: 0.1 - 3.8 N/m	environment: air
force-sensing tip: shredded diamond	mode: dc
sample: HOPG, LiF, carbon films	force: atomic attractive and repulsive

Table 4.7 Atomic force microscopy (Sugawara et al. 1990a,b).

Lever and tip specifications	Performance highlights
lever material: W	typical force: 2.2×10^{-10} N
lever coating: Au	spatial resolution: atomic
lever shape: rectangular	environment: air
lever length: 350 - 800 μm	mode: dc
lever width: 200 μm	force: atomic attractive and repulsive
lever thickness: 3 μm	
lever spring constant: 1 - 11 N/m	
lever resonance frequency: 3.5 - 16.2 kHz	
force-sensing tip: Au or W	
tunneling tip: Pt-Ir	
sample: HOPG	

Table 4.8 Atomic force microscopy (Marti et al. 1987 and Marti et al. 1988a,b,c,d).

Lever and tip specifications	Performance highlights
lever material: Pt	typical force: 10^{-8} N
lever shape: double cross (*V*)	spatial resolution: atomic (HOPG,Si)
cross distance: 5 mm	environment: air, oil
lever length: 4 mm	mode: dc
lever diameter: 25 μm	force: atomic repulsive
lever spring constant: 40 N/m (Pt)	
lever resonance frequency: 3 kHz	
force-sensing tip: shredded 0.5-mm diamond	
tunneling tip: triangular wire	
sample: HOPG and NaCl under paraffin oil, Si	

Table 4.9 Atomic force microscopy (Yang et al. 1988, Bryant et al. 1988a,b, and Bryant et al. 1989).

Lever and tip specifications	Performance highlights
lever material: Pt, Pt-Ir, W, C, Au	typical force: 10^{-7} N
lever shape: bent wire	environment: air
lever spring constant: 0.1 - 200 N/m	mode: dc
lever resonance frequency: 10.2-100 kHz	force: atomic repulsive
force-sensing tip: same material	
sample: HOPG, Ag-doped CdTe, Au-coated mica	

Table 4.10 Atomic force microscopy (Burnham and Colton 1989 and Burnham et al. 1990).

Lever and tip specifications	Performance highlights
lever material: Au-coated tungsten	typical force: 10^{-8} N
lever shape: two bent wires	typical lever deflection: 0.2 Å
lever diameter: 20 μm	environment: air
lever spring constant: 50 N/m	mode: dc
force-sensing tip: tungsten	force: atomic attractive and adhesive
sample: HOPG, Au, elastomer, PTFE, Al LB layers	

Table 4.11 Atomic force microscopy (Dürig et al. 1986).

Lever and tip specifications	Performance highlights
lever material: steel (copper, Ta)	typical force: 10^{-9} N
lever shape: rectangular	thermal excitation: 0.25 Å
lever length: 10 mm (5.8 mm)	spatial resolution: 10 Å
lever width: 1.5 mm (1.15 mm)	environment: UHV
lever thickness: 50 μm (20 μm)	mode: spectrum
lever spring constant: 10 N/m (1 N/m)	force: atomic repulsive
lever resonance frequency: 415 Hz (280 Hz)	
tunneling and force-sensing tip: W	
sample: Ag	

Table 4.12 Atomic force microscopy (Dürig and Züger 1989).

Lever and tip specifications	Performance highlights
lever material: Ir	amplitude of vibration: 0.3 Å*
lever shape: rectangular	typical force: 10^{-9} N
lever length: 265 μm (349 μm)	typical lever deflection: 0.5 Å
lever width: 0.45 mm	environment: UHV
lever thickness: 50 μm	mode: spectrum
lever spring constant: 400 N/m (170 N/m)	force: atomic repulsive
lever resonance frequency: 1.393 kHz	
lever quality factor: 400	
tunneling and force-sensing tip: Ir	
sample: Ir, Au, Al	*(white noise)

Table 4.13 Atomic force microscopy (Tang et al. 1988).

Lever and tip specifications	Performance highlights
lever spring constant: 8×10^{4} N/m	typical force: 2×10^{-6} N
lever resonance frequency: 1.4 Hz	lever deflection: 0.1 Å
tunneling and force-sensing tip: tungsten	environment: air
sample: HOPG	mode: spectrum
	force: atomic repulsive

4.4. Cross Arrangement

Figure 4.1(b) shows a schematic of the system where the lever consists of crossed wires supporting the force-sensing tip.

4.5. Parallel Arrangement

Figure 4.1(c) shows a schematic of the AFM-STM system where the lever supporting the force-sensing tip is attached parallel to the tunneling tip and has a bend at its end. Stabilized STM-tip and AFM-lever operation were achieved using a liquid between the two (Bryant et al. 1990 and Cheng et al. 1990).

4.6. Serial Arrangement

Figure 4.1(d) shows a schematic of the system where the tunneling tip is in series with the force-sensing tip, which is supported by two bent wires.

4.7. Single-Lever Arrangement

Figure 4.1(e) shows a schematic of the system where the sample is mounted on a lever and the tunneling tip also acts as the force-sensing tip. For other configurations, see Moreland and Rice (1990a, 1991), DiCarlo et al. (1992), and Yi et al. (1992).

4.8. Summary

We have seen that in a tunneling detection system, the tunneling tip is located several angstroms from the lever supporting the force-sensing tip, that the interaction is between a single atom at the apex of the tip and its closest atom on the lever, and that strong atomic forces act between these two. The tunneling current density j, used for tracking and positioning the lever, can be approximated by

$$j = \frac{e^2}{\hbar} \frac{\kappa_0}{4\pi^2 z} V \exp(-2\kappa_0 z) ,$$

which is exponential in the tunneling gap.

5
Capacitance Detection System

5.1. Introduction

Figure 5.1 is a schematic of a capacitance detection system where the minute deflections of a lever supporting a force-sensing tip are monitored by the capacitance of the lever and a reference plate. This chapter reviews four implementations of such a system, all of which use a high-Q-tuned electronic circuit to measure the small changes in the capacitance. One can operate on the slope of the electronic resonance curve of the circuit, where small resonance frequency changes caused by changes in the capacitance translate into voltage changes that are used to map the capacitance across the sample. To enhance the sensitivity of the system, the lever can be vibrated on the slope of its mechanical resonance curve, so that small changes in the force derivative acting on the lever vary its amplitude of vibration. Consequently, the oscillating voltage of the electronically tuned circuit will be amplitude modulated by the vibration of the lever, making it possible to map the force derivative across a sample.

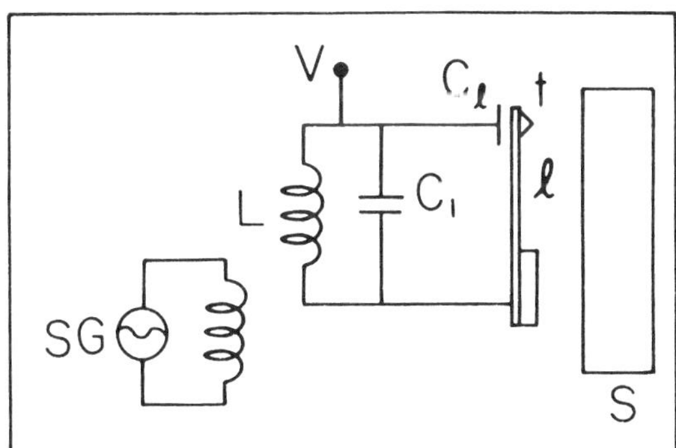

Fig. 5.1 Schematic of the capacitance detection system showing the signal generator SG exciting a high-Q-tuned circuit composed of an inductance L, and capacitances C_1 and C_P, and the output voltage V.

5.2. Theory

The capacitance in the scanning force microscope is determined by the circuit and stray capacitance and the capacitor that consists of the plate supported by the lever and reference plate, denoted by C_ℓ. The voltage across the tuned circuit, as a function of frequency, is given by

$$V = \frac{V_0}{\sqrt{1 + 4Q^2 \, \Delta f^2 / f_0^2}} \,. \tag{5.1}$$

Here $\Delta f = |f - f_0|$, and V_0 is the voltage at the resonance frequency f_0 given by

$$f_0 = \frac{1}{2\pi} \sqrt{LC} \,, \tag{5.2}$$

where Q is the quality factor and L is the inductance. The bandwidth B of the circuit is obtained from

$$B = 2|f_{3dB} - f_0| \,, \tag{5.3}$$

where f_{3dB} is that frequency for which the amplitude decreases by a factor of $\sqrt{2}$ from its maximum value, or the power decreases by 3 dB. Q is related to the bandwidth B by

$$Q = f_0/B \,. \tag{5.4}$$

Maintaining a constant operating frequency f will change the voltage V when the resonance frequency f_0 shifts because of changes in the capacitance C. We want to determine the sensitivity of the tuned circuit to changes in the capacitance C for operation on the slope of the resonance curve of the circuit. To that end, we write the change in the voltage, δV, in terms of the two derivatives dV/df and df/dC:

$$\delta V = \frac{dV}{df} \frac{df}{dC} \delta C \,. \tag{5.5}$$

These derivatives are given by (Palmer et al. 1982)

$$\frac{dV}{df} = \alpha \frac{V_0}{B} \tag{5.6}$$

and

$$\frac{df}{dC} = \gamma \frac{f}{C} , \tag{5.7}$$

where $\alpha = \sqrt{3}/4$ for $f = f_{6dB}$, and $\gamma = -0.5$ for a single-tuned LC circuit. Inserting Eqs. (5.6) and (5.7) into Eq. (5.5) yields

$$\delta V = \alpha \frac{\gamma}{C} V_0 \frac{f}{B} \delta C . \tag{5.8}$$

We find that the sensitivity increases if we use a small capacitance C and operate at a high frequency. The capacitors comprising C are in parallel and give a total capacitance of

$$C = C_1 + C_\ell . \tag{5.9}$$

The capacitance C_ℓ, in turn, can be approximated by modeling it as that of a plane-parallel capacitor, given by

$$C_\ell = \epsilon_0 \frac{w\ell}{z} , \tag{5.10}$$

where ϵ_0 is the permittivity of free space, ℓ and w are the length and width of each of the two plates, and z is their separation. Assume now that the lever vibrates around its equilibrium position, z_0, with amplitude A and frequency $f_\ell = \Omega/2\pi << f$, yielding

$$z = z_0 + A \sin(\Omega t) . \tag{5.11}$$

Since $A << z_0$, we can invert Eq. (5.11) and obtain

$$\frac{1}{z} = \frac{1}{z_0} \left[1 - \frac{A}{z_0} \sin(\Omega t) \right] , \tag{5.12}$$

which can be inserted into Eq. (5.10) yielding

$$C_\ell = \epsilon_0 \frac{w\ell}{z_0} \left[1 - \frac{A}{z_0} \sin(\Omega t) \right] . \tag{5.13}$$

We now decompose C_ℓ into two components,

$$C_\ell = C_{\ell_0} + \delta C_\ell \tag{5.14}$$

with

$$C_{\ell_0} = \epsilon_0 \frac{w\ell}{z_0} \tag{5.15}$$

and

$$\delta C_{\ell} = \epsilon_0 \frac{w\ell}{z_0^2} \mathrm{A}\sin(\Omega t) \ . \tag{5.16}$$

Inserting Eq. (5.13) into Eq. (5.8) finally yields

$$i_{\Omega} = \beta \mathrm{A} \sin(\Omega t) \ , \tag{5.17}$$

where

$$\beta = \alpha\gamma \frac{i_0}{C} \frac{f}{B} \epsilon_0 \frac{w\ell}{z_0^2} \ , \tag{5.18}$$

and where we have replaced the voltages by their respective currents using the constant β. We find, therefore, that the voltage across the tuned circuit, which oscillates at frequency f, will be amplitude modulated by the vibrating lever at frequency Ω, and measuring the amplitude of this modulation gives the amplitude of vibration of the lever. Two-dimensional deflection methods have been discussed by Cohen et al. (1990a,b) and Neubauer et al. (1990) (see also, Grigg et al. 1992b, Bridges and Thomson 1992, Bergasa and Sáenz 1992, Huang et al. 1992, and Barrett and Quate 1991).

5.3. Noise Considerations

Johnson and shot noise do not exist in the capacitance detection system.

5.3.1. Capacitance Drift

The capacitance detection system suffers from effects of drift in the tip-sample capacitance and from stray capacitance. Since the relative thermal drift $\delta z/z_0 \propto \delta T$, where δT is the temperature change, the drift can be minimized by using a small tip-sample distance, z_0.

5.3.2. Oscillator Noise

A general expression of the oscillator noise can be written as

$$\langle \delta i^2 \rangle_O = \text{RIN } i^2 . \tag{5.19}$$

5.3.3. Lever Thermal Noise

On resonance, we found that the lever noise is

$$\langle \delta A^2 \rangle_\ell = \frac{4KTBQ}{\omega_0 k} , \tag{5.20}$$

which translates into a current noise of the lever

$$\langle \delta i^2 \rangle_\ell = \beta^2 \frac{4KTBQ}{\omega_0 k} . \tag{5.21}$$

For off-resonance operation we get

$$\langle \delta i^2 \rangle_\ell = \beta^2 \frac{4KTB}{Q\omega_0 k} . \tag{5.22}$$

5.3.4. Signal-to-Noise Ratio

The total signal-to-noise ratio for on- and off-resonance operation is

$$\frac{i}{\langle \delta i^2 \rangle^{1/2}} = \beta \frac{A}{\sqrt{\langle i \rangle^2 \text{ RIN} + \beta^2 \dfrac{4KTBQ}{\omega_0 k}}} \tag{5.23}$$

and

$$\frac{i}{\langle \delta i^2 \rangle^{1/2}} = \beta \frac{A}{\sqrt{\langle i \rangle^2 \text{ RIN} + \beta^2 \dfrac{4KTB}{Q\omega_0 k}}} . \tag{5.24}$$

5.4. Performance of Systems

5.4.1. Scanning Capacitance Microscopy (Matey and Blanc 1985)

The scanning capacitance microscope is based on an RCA VideoDisc (Palmer et al. 1982) that sensed capacitively the 80-Å-high bits on a video disk. The system measured the capacitance between a probe and a sample, rather than between a probe and a force-sensing lever.

Consequently, it does not provide force information. The sensor in the experiments was a thin electrode deposited on a diamond stylus that was dragged across the disk with 4-μm-deep, 2.6-μm-pitch spiral grooves. The electrode can be modeled as an infinitely thin conducting sheet placed at a right angle to a conducting sample. The change in the capacitance of such a structure is

$$\frac{dC}{dz} = \frac{8}{\pi} \epsilon_0 \epsilon_r \frac{w}{z} , \tag{5.25}$$

where w is the width of the electrode and z is its distance from the conducting sample. Note that the thickness of the electrode plays no role here, as the electric field lines emanate from the sides. To measure such a small capacitance change, the authors employed an ultrahigh frequency oscillator that excited a tuned circuit consisting of an inductor in parallel to two capacitors. Table 5.1 illustrates sensor specifications and performance highlights.

Table 5.1 Scanning capacitance microscopy (Matey and Blanc 1985).

Lever and tip specifications	Performance highlights
material: diamond stylus	$\partial C / \partial z$ =3×10^{-9} F/m
dimensions: 5 μm × 2.5 μm	height resolution: 3 Å
metal electrode thickness: w =0.15 μm	lateral resolution: 1000 Å
electronic circuit Q: 30	oscillator noise: 1 μV
stray capacitance: 0.1 pF	(B =30 kHz at 5 MHz)

5.4.2. Scanning Capacitance Microscopy (Williams et al. 1989)

The scanning capacitance microscope is an upgraded version of that of Matey and Blanc (1985), which employed a fine lever mounted on a piezoelectric tube. The system utilized a scanning-tunneling-microscope feedback system to control the tip-sample distance, and the tip was vibrated to reduce excessive noise. The sensitivity of their system, assuming a "square" tip apex, is obtained from the derivative of the capacitance. For example, a 100-Å × 100-Å "square" tip at a distance of 20 Å gives C = 4.4x10^{-19} F and $\partial C/\partial z$ = 4.4x10^{-11} F/m. The samples were two-dimensional arrays of 500-Å-diameter circular holes on 1000-Å centers fabricated in 1000-Å-thick polymethyl-methacrylate (PMMA) with a 200-Å-thick gold coating. The authors conclude that their system is capable of a resolution better than 100 Å (see Table 5.2).

Table 5.2 Scanning capacitance microscopy (Williams et al. 1989).

Lever and tip specifications	Performance highlights
material: tungsten	sensitivity: 70 mV/ fF
shape: rod	lateral resolution: 250 Å
diameter: 0.01 in.	minimum capacitance: 2×10^{-22} F
tip diameter: 250 Å	minimum charge: 0.01 electrons
resonance frequency: 72 kHz	noise: 7 μV (B = 1 kHz at 10 kHz)
sample: PMMA with a coating of 200 Å gold	
electronic resonance frequency: 915 MHz	

5.4.3. Magnetic Force Microscope (Göddenhenrich et al. 1990)

The system employed a force-sensing tip that was attached to one plate of a capacitor while the other plate was attached to a single piezoelectric tube. This arrangement has an advantage relative to the two previously mentioned systems in that the control capacitance is larger than the tip-sample capacitance, and as a result, the effect of stray capacitance is reduced. To measure the capacitance, they used a capacitance transformer bridge with an arms ratio of 100:1. Since the control capacitor was approximately 1 pF, the reference capacitor could be as large as 100 pF. The sensitivity was limited by the drift of the reference capacitor. The system could be operated in two different modes, depending on whether or not the tip-to-sample distance was modulated. The system could image 150×2-μm^2 magnetic bits on 3.5-in. flexible disks. With nonmagnetic tips, it could image the topography of a disk that had clusters with several-thousand-Å width and several-hundred-Å depth. Table 5.3 lists specifications and performance highlights of this system.

Table 5.3 Magnetic force microscopy (Göddenhenrich et al. 1990).

Lever and tip specifications	Performance highlights
material: Al	minimum force: 10^{-10} N
shape: rod	lateral resolution: >1000 Å
diameter: 25 μm	minimum capacitance: 10^{-19} F
tip material: Ni	
spring constant: 1 N/m	
capacitor material: Al foil	
capacitor dimensions: 300 μm × 300 μm	
sample: magnetic media	

5.4.4. Scanning Force Microscope (Neubauer et al. 1990 and Cohen et al. 1990a,b)

The scanning force microscope, operating under UHV conditions, uses a dual-capacitance sensing method for simultaneous measurements of normal and lateral forces. The instrument, which is aimed at measuring the tribological properties of materials such as friction, lubrication, and wear, exhibits a high sensitivity and dynamic range, while maintaining a low noise level at low frequencies. The design of the levers took into account the snapping distance at which the electrostatic force derivative equals the spring constant of the lever. The electronic circuit, which is described in detail in the references, does not have a resonance circuit, and the chief source of noise was the variations in stray capacitance.

Table 5.4 Scanning force microscopy (Neubauer et al. 1990 and Cohen et al. 1990a,b).

Lever and tip specifications	Performance highlights
material: Ir	max. capacitance derivative: 5×10^{-7} F/m
shape: bent rod	lateral resolutions: 30 Å
diameter: 125 μm	noise: 0.02 Å (0.1 - 10 kHz)
length: 6 mm	height resolution: 10 Å
tip material: Ir (diamond)	
spring constant: 90 N/m	
sample: HOPG (diamond film)	

5.5. Summary

The current generated by the capacitance detection system is

$$i = \alpha\gamma \frac{i_0}{C} \frac{f_0}{B} \epsilon_0 \frac{w\ell}{z_0^2} A \sin(\Omega t) ,$$

where $\alpha\gamma = \sqrt{3}/8$, C is the lever-reference plate capacitance whose separation is z_0, f_0 is the resonance frequency and bandwidth of the tuned circuit, ϵ_0 is the free-space permittivity, w and ℓ are the width and length of the lever, and A and Ω are the amplitude and frequency of vibration of the lever.

6
Homodyne Detection System

6.1. Introduction

In a basic homodyne detection system, shown schematically in Fig. 6.1, a polarized laser beam passes through a beam splitter and is incident on a Fabry Perot whose reflecting surfaces consist of an optical flat and a lever supporting the force-sensing tip. The beam reflected back from the Fabry Perot is incident on the same beam splitter and is deflected into a photodetector that generates a photocurrent used to image the force acting on the tip. In a differential homodyne detection system, a fraction of the laser power, serving as a reference beam, is diverted by a beam splitter to a first photodetector. The light passing through the beam splitter, acting as the signal beam, is incident on the Fabry Perot, reflected back, and deflected into a second photodiode. The currents of the two photodetectors are compared, and the difference yields a signal that is used to image the force acting on the tip. Actual systems will contain many more components that will improve their operation, as will be discussed later. If the lever oscillates at frequency Ω, then the photocurrent will consist of components with frequency 0, Ω, and 2Ω, to the second order. Using a phase-sensitive detector we can obtain the amplitude of vibration from the ratio of the first and second harmonics.

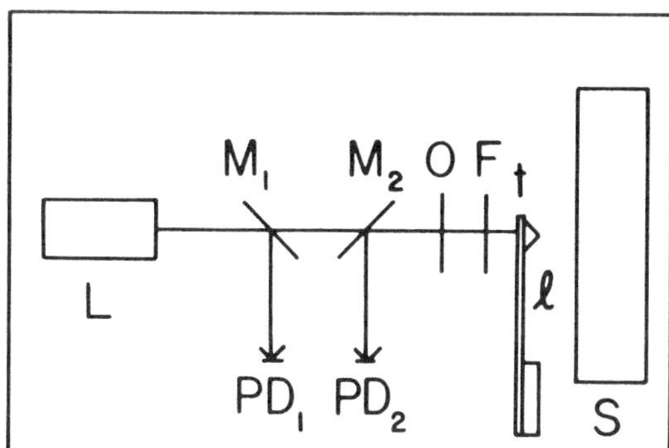

Fig. 6.1 A schematic diagram of the homodyne detection system showing the laser (L), the beam splitters μ_1 and μ_2, the lever (ℓ) supporting the force-sensing tip (t), the sample (S), the microscope objective (O), the optical flat (F), and the photodetectors PD_1 and PD_2.

6.2. Theory

A homodyne detection process can be analyzed by calculating the interference between the various waves, as shown in Fig. 6.2. The figure shows the reflection and transmission of a wave incident on the optical flat, which is separated from the lever by a distance z. Let r and t be the reflection and transmission coefficients of the optical flat and r' and t' those of the lever, respectively, and let E^i, E^r, and E^t be the amplitude of the incident, reflected, and transmitted waves, respectively. Using this notation, we get

$$E_r = rE_i + tt'r'\exp(i\theta)E_i + tt'r'^3\exp(2i\theta)E_i + \dots \tag{6.1}$$

and

$$E_t = tt'E_i + tt'r'^2\exp(2i\theta)E_i + tt'r'^4\exp(2i\theta)E_i + \dots \tag{6.2}$$

for the reflected and transmitted waves. Here $E_r = E_{r_0}\sin(\omega t)$ and $E_t = E_{t_0}\sin(\omega t)$, where ω is the optical frequency and

$$\theta = \frac{4\pi}{\lambda} z \ . \tag{6.3}$$

By summing the infinite series of Eqs. (6.1) and (6.2) we obtain for the reflected and transmitted power, respectively,

$$P_r = P_i \frac{4R\sin^2(\theta/2)}{(1 - R)^2 + 4R\sin^2(\theta/2)} \tag{6.4}$$

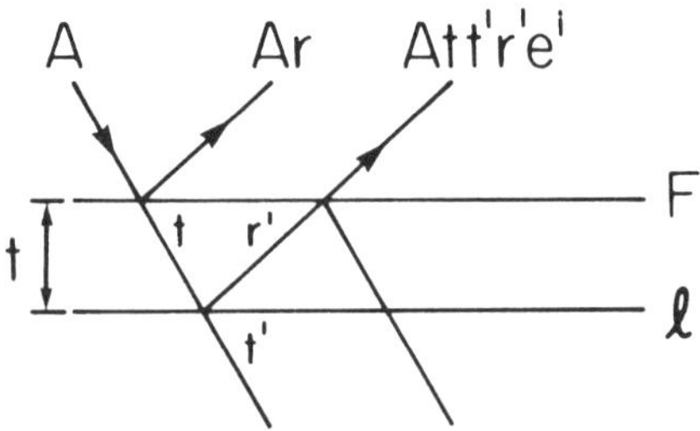

Fig. 6.2 The optical waves in a Fabry-Perot interferometer.

and

$$P_t = P_i \frac{T^2}{(1 - R)^2 + 4R\sin^2(\theta/2)}, \tag{6.5}$$

where P_i is the incident power and

$$R = rr', \quad T = tt' . \tag{6.6}$$

Figure 6.3 shows P_r/P_i as a function of z/λ for $R = 0.1$, 0.2, and 0.3, where we observe that as the resonances become sharper, their amplitude increases. The lever reflectivity is usually high while the optical-flat reflectivity is low, resulting in a small effective R. Under these conditions we can approximate Eq. (6.4) and write the power incident on the right-hand-side photodiode, P_{pd2}, which we assume to be equal to P_r:

$$P_{pd2} = \frac{1}{2} FP_i \left[1 - \cos\left(\frac{4\pi z}{\lambda}\right)\right], \tag{6.7}$$

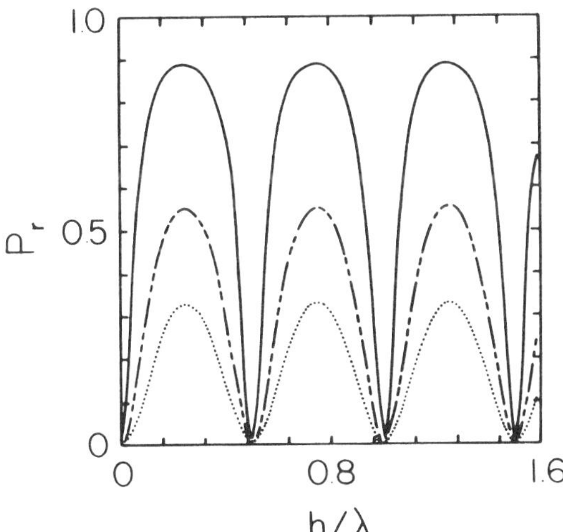

Fig. 6.3 The scaled reflected power P_r/P_i as a function of z/λ for $R = 0.1$, 0.2, and 0.3; z is the lever-optical flat distance and λ is the optical wavelength.

where

$$F = \frac{4R}{(1 - R)^2} . \tag{6.8}$$

When the lever vibrates with amplitude A and frequency Ω,

$$z = z_0 + A\sin(\Omega t) , \tag{6.9}$$

and we obtain

$$P_{pd2} = \frac{1}{2} FP_i \left[1 - \cos\left[\frac{4\pi}{\lambda}[z_0 + A\sin(\Omega t)]\right]\right] . \tag{6.10}$$

We can expand the sine and cosine functions assuming that

$$\frac{4\pi}{\lambda} A << 1 \tag{6.11}$$

and obtain the reflected power at dc and at the first and the second harmonics:

$$P_{pd2,dc} = \frac{1}{2} FP_i \left[1 - \cos\left(\frac{4\pi z_0}{\lambda}\right) + \frac{1}{4}\left(\frac{4\pi A}{\lambda}\right)^2 \cos\left(\frac{4\pi z_0}{\lambda}\right)\right] , \tag{6.12}$$

$$P_{pd2,\Omega} = \frac{1}{2} FP_i \frac{4\pi}{\lambda} A \sin\left(\frac{4\pi z_0}{\lambda}\right) \sin(\Omega t) , \tag{6.13}$$

and

$$P_{pd2,2\Omega} = -\frac{1}{8} FP_i \left(\frac{4\pi A}{\lambda}\right)^2 \cos\left(\frac{4\pi z_0}{\lambda}\right) \cos(2\Omega t) . \tag{6.14}$$

We now adjust the distance z_0 such that

$$\cos\left(\frac{4\pi}{\lambda} z_0\right) = 0 , \tag{6.15}$$

where the optical power incident on the photodetector will generate a photocurrent

$$i = \eta P_{pd2} \, , \tag{6.16}$$

and η is the quantum efficiency of the photodetector. The photodetector current at the frequency Ω, from Eq. (6.10), is

$$i_\Omega = \frac{1}{2} \eta F P_i \frac{4\pi}{\lambda} A \sin(\Omega t) \, , \tag{6.17}$$

which is proportional to the amplitude of vibration, while the component at the frequency 2Ω is zero. The quadrature adjustment generates therefore a photocurrent oscillating around the dc value

$$i_0 = \frac{1}{2} \eta F P_i \, . \tag{6.18}$$

If we adjust the distance z_0 such that

$$\cos\left(\frac{4\pi z_0}{\lambda}\right) = \pm 1 \, , \tag{6.19}$$

we obtain a photocurrent at the frequency 2Ω given by

$$i_{2\Omega} = \frac{1}{8} \eta F P_i \left(\frac{4\pi}{\lambda} A\right)^2 \cos(2\Omega t) \, , \tag{6.20}$$

which is proportional to the square of the amplitude of vibration, while the photocurrent at frequency Ω is now zero. The oscillating photocurrent at frequency 2Ω is around a dc value given by either

$$i_0 = \eta F P_i \tag{6.21}$$

or

$$i_0 \simeq 0 \, , \tag{6.22}$$

depending on whether $\cos(4\pi z_0/\lambda)$ equals -1 or 1. The amplitude of vibration can be obtained from the ratio of the second to first harmonics:

$$A = \frac{\lambda}{\pi} \frac{i_{2\Omega}}{i_\Omega} \, . \tag{6.23}$$

Figure 6.4 shows the first and second harmonics obtained for a modulation of the lever at quadrature (B) and at the extrema of the reflectivity curve (A,C). The output of the photodetector is conveniently probed by a phase-sensitive detector that can be adjusted to measure the sine or cosine components of the first or second harmonics. A drawback of the homodyne technique is that the signal depends on the parameter z_0, which can drift because of thermal or mechanical instabilities and cause the signal from the phase-sensitive detector to drift. To compensate for this effect, we can use two phase-sensitive detectors tuned to the first and second harmonics, respectively. The first phase-sensitive detector will generate a signal that is proportional to the lever amplitude of vibration, which is used to image the forces. The second lock-in amplifier will generate a signal that drives a feedback circuit, which adjusts the distance z_0 so that the second harmonic remains at zero.

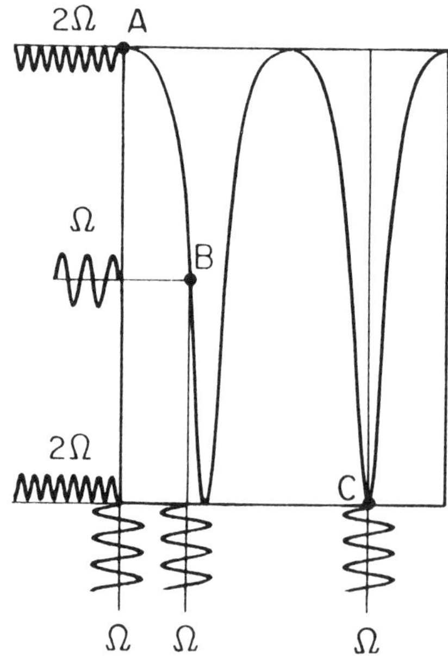

Fig. 6.4 Generation of the first and second harmonics by modulating the lever at quadrature and at the extrema of the reflectivity curve, respectively. The horizontal and vertical axes relate to the lever displacement and interferometer output, respectively.

6.3. Noise Considerations

6.3.1. Basic Homodyne Detection System

The basic homodyne detection system uses only the second photodetector, PD_2, to drive a current amplifier.

6.3.1.1. Optical pathlength drift. The homodyne detection system suffers from effects of drift in the optical pathlength z_0. Since the relative thermal drift $\delta z/z_0 \propto \delta T$, where δT is the temperature change, we can minimize δz by using z_0 as small as 10 μm.

6.3.1.2. Johnson noise. The Johnson noise generated by the detector is

$$\langle \delta i^2 \rangle_J = 4KTB/R \tag{6.24}$$

and, in comparison to the other noise contributions, can be neglected.

6.3.1.3. Laser noise. Using Eq. (3.24) we find that the current noise of the laser

$$\langle \delta i^2 \rangle_L = \frac{1}{4} \eta^2 F^2 P^2{}_i \text{ RIN} . \tag{6.25}$$

6.3.1.4. Lever thermal noise. On resonance, we found that the lever thermal noise translates into a current noise caused by the lever

$$\langle \delta i^2 \rangle_\ell = \frac{1}{4} \eta^2 F^2 P^2{}_i \frac{16\pi^2}{\lambda^2} \frac{4KTBQ}{\omega_0 k} , \tag{6.26}$$

and off resonance, we get

$$\langle \delta i^2 \rangle_\ell = \frac{1}{4} \eta^2 F^2 P^2{}_i \frac{16\pi^2}{\lambda^2} \frac{4KTB}{Q\omega_0 k} \tag{6.27}$$

for the current noise caused by the lever.

6.3.1.5. Shot noise. The shot noise generated by the photodetector PD_2, using Eq. (3.16), is

$$\langle \delta i^2 \rangle_s = e\eta F P_i B . \tag{6.28}$$

6.3.1.6. Signal-to-noise ratio. Neglecting the Johnson noise, which is the smallest contribution, the signal-to-noise ratios attributable to the other noise contributions for on- and off-resonance operation are, respectively,

$$\frac{i}{\langle \delta i^2 \rangle^{1/2}} = \frac{4\pi}{\lambda} \frac{A}{\sqrt{\mathrm{RIN} + \dfrac{16\pi^2}{\lambda^2}\dfrac{4KTBQ}{\omega_0 k} + \dfrac{4eB}{\eta F P_i}}} \tag{6.29}$$

and

$$\frac{i}{\langle \delta i^2 \rangle^{1/2}} = \frac{4\pi}{\lambda} \frac{A}{\sqrt{\mathrm{RIN} + \dfrac{16\pi^2}{\lambda^2}\dfrac{4KTB}{Q\omega_0 k} + \dfrac{4eB}{\eta F P_i}}} \,. \tag{6.30}$$

6.3.2. Differential Homodyne Detection System

The differential homodyne detection system uses both photodetectors, PD_1 and PD_2, to drive a difference-current amplifier. This amplifier can be operated at low frequencies, almost at dc, because the use of both photodetectors practically cancels out laser noise.

6.3.2.1. Optical pathlength drift. The differential homodyne detection system suffers from the same effects of drift in the optical path length z_0 as the basic system. These effects can be minimized by choosing z_0 to be as small as 10 μm.

6.3.2.2. Johnson noise. The Johnson noise generated by each detector is

$$\langle \delta i_1{}^2 \rangle_J = 4KTB/R \,, \tag{6.31}$$

and the noise fed into the difference amplifier is

$$\langle \delta i^2 \rangle_J = 8KTB/R \,, \tag{6.32}$$

which is larger by a factor of 2 than the noise of each detector. In comparison to the other noise contributions, however, it can be neglected.

6.3.2.3. Laser noise. The current noise of the laser generated by each photodiode is

$$\langle \delta i_1{}^2 \rangle_L = \eta^2 P^2{}_{pd1} \, \mathrm{RIN} \,. \tag{6.33}$$

The current noise of the laser in the photodetector pd_2 is

$$\langle \delta i_2{}^2 \rangle_L = \eta^2 P^2{}_{pd2}\ \mathrm{RIN}\ . \tag{6.34}$$

The low-frequency noise at the output of the differential amplifier is obtained by subtracting Eq. (6.34) from Eq. (6.33). For a balanced system, where $P_{pd1} = P_{pd2}$, the laser noise can be neglected in comparison to the other sources of noise.

6.3.2.4. Lever thermal noise. On resonance, we found that the thermal noise of the lever is

$$\langle \delta A^2 \rangle_\ell = \frac{4KTBQ}{\omega_0 k}\ , \tag{6.35}$$

which in this case contributes noise to only one of the detectors. The current noise is therefore

$$\langle \delta i^2 \rangle_\ell = \frac{1}{4}\ \eta^2 F^2 P^2{}_i\ \frac{16\pi^2}{\lambda^2}\ \frac{4KTBQ}{\omega_0 k}\ . \tag{6.36}$$

6.3.2.5. Shot noise. The shot noise generated by each detector is

$$\langle \delta i_1{}^2 \rangle_s = 2e\eta P_{pd1} B \tag{6.37}$$

and

$$\langle \delta i_2{}^2 \rangle_s = 2e\eta P_{pd2} B\ , \tag{6.38}$$

and the noise fed into the difference amplifier is then

$$\langle \delta i^2 \rangle_s = 2e\eta (P_{pd1} + P_{pd2}) B\ . \tag{6.39}$$

6.3.2.6. Signal-to-noise ratio. Neglecting Johnson and laser noise, which are the smallest contributions now, the signal-to-noise ratio attributable to the other contributions for on- and off-resonance operation are, respectively,

$$\frac{i}{\langle \delta i^2 \rangle^{1/2}} = \frac{4\pi}{\lambda}\ \frac{A}{\sqrt{RIN_{in} + \dfrac{16\pi^2}{\lambda^2}\dfrac{4KTBQ}{\omega_0 k} + \dfrac{8eB}{\eta(P_{pd1} + P_{pd2})}}} \tag{6.40}$$

and

$$\frac{i}{\langle \delta i^2 \rangle^{1/2}} = \frac{4\pi}{\lambda} \frac{A}{\sqrt{RIN_{in} + \frac{16\pi^2}{\lambda^2}\frac{4KTB}{Q\omega_0 k} + \frac{8eB}{\eta(P_{pd1} + P_{pd2})}}}. \tag{6.41}$$

6.4. System Performance

6.4.1. Basic Homodyne Detection System

McClelland et al. (1987), Mate et al. (1987), Erlandsson et al. (1988a), and Mate et al. (1988) employed the system shown schematically in Fig. 6.1 without PD_1. Here a polarized beam from a HeNe laser is incident on a polarizing beam splitter. Because of the orientation of the polarization, the polarizing beam splitter will deflect the beam 90^0 into the quarter-wave plate. The quarter-wave plate introduces a phase shift between the ordinary and extraordinary vibrations. When it is oriented 45^0 with respect to the plane of polarization of the light, it will change the beam from linear to circular. The beam continues toward a flat half mirror and part of it is reflected back and passes again through the quarter-wave plate that converts its polarization to linear and rotates it 90^0. This first beam, whose polarization has been rotated, will now be deflected by the polarizing beam splitter into the photodiode. The second part of the beam passes through the flat half mirror and is focused onto the lever supporting the force-sensing tip and is reflected back. The beam then traverses the same path as the first beam and is incident on the same photodiode. The interference between these two beams is affected by the optical pathlength difference between the flat half mirror and the lever. The minute deflections of the lever, under the influence of the force field, are detected in "quadrature," namely under the condition that the average phase shift between the two beams is off by 90^0. If the lever is vibrated at a frequency Ω, then the photodetector will generate a current at this frequency. The system could be operated in the dc mode where the lever was not vibrated, and in the ac mode where the lever was vibrated by a bimorph and its resonance frequency monitored by the homodyne detection system. Alternatively, the sample was vibrated and the lever amplitude of vibration was monitored. Tables 6.1 through 6.4 detail lever and tip specifications and their performance.

Table 6.1 Atomic force microscopy (McClelland et al. 1987).

Lever and tip specifications	Performance highlights
material: tungsten shape: bent rod length: 3 mm base diameter: 70 μm	amplitude of vibration: 1 Å smallest detectable force: 3×10^{-9} N spatial resolution: 200 Å

Table 6.2 Atomic force microscopy (Mate et al. 1987).

Lever and tip specifications	Performance highlights
material: tungsten shape: bent rod length: 12 mm base diameter: 0.25 and 0.5 mm tip diameter: 1500 - 3000 Å spring constant: 150 and 2500 N/m	smallest detectable repulsive force: 2×10^{-7} N spatial resolution: atomic

Table 6.3 Atomic force microscopy (Erlandsson et al. 1988a).

Lever and tip specifications	Performance highlights
material: tungsten shape: bent rod length: 3 mm base diameter: 70 μm tip diameter: 1500 - 3000 Å spring constant: 30 N/m resonance frequency: 5 kHz quality factor: 300	amplitude of vibration: 2 Å smallest detectable force: 3×10^{-10} N smallest lever deflection: 0.1 Å spatial resolution: 50 Å noise: 0.2 Å for 0.5 - 3 kHz noise: 0.05 Å for 100 Hz - 1 kHz thermal excitation: 0.1 Å

Table 6.4 Atomic force microscopy (Mate et al. 1987, 1988).

Lever and tip specifications	Performance highlights
material: Pt-Rh, W shape: bent rod base diameter: 0.25 and 0.5 mm spring constant: 150 → 2500 N/m	spatial resolution: atomic

6.4.2. Fiber-Coupled Homodyne Detection

Stern et al. (1988), Mamin et al. (1988), and Rugar et al. (1989) employed the system shown schematically in Fig. 6.1 using a fiber-coupled laser (see also, Mulhern et al. 1991, Watanabe et al. 1992, and Grütter et al. 1992). Here a polarized beam from a HeNe laser passes through a Faraday isolator that protects the laser from reflections, a beam expander, and a half-wave plate that rotates its polarization. The Faraday isolator is a magnetically activated medium, usually quartz, that rotates the polarization plane of the light at an angle proportional to the magnetic field and the thickness of the medium. When a plane-polarized beam passes back and forth through the Faraday isolator that protects the laser from reflections, its rotation will further increase, in contrast to a natural polarization rotator for which this rotation will cancel. The beam continues and passes through a polarizing beam splitter and a microscope objective that focuses it onto a single-mode fiber with a core diameter of 4 μm. The fiber is bent into a 360^0 loop with a 2-cm radius, and the induced strain makes the fiber act as a quarter-wave plate. The other side of the fiber is placed within a distance, 10 μm, from the lever supporting the force-sensing tip. The front face of the fiber and the lever constitute a Fabry Perot with thickness ℓ. The back reflection of the Fabry Perot retraces its original path until it arrives at the polarizing beam splitter, which reflects it into a photodetector. In comparison to the previous method, the optical pathlength difference here, which is the thickness of the Fabry Perot, is smaller, and the system is therefore more stable. Tables 6.5 through 6.7 detail lever and tip specifications and performance.

Table 6.5 Magnetic force microscopy (Rugar et al. 1989).

Lever and tip specifications	Performance highlights
material: iron	amplitude of vibration: 50 Å
shape: rod	spatial resolution: several μm
length: 550 μm	noise amplitude: 1.7×10^{-4} Å/$\sqrt{\text{Hz}}$ above 2 kHz
base diameter: 15 μm	noise amplitude: 0.01 Å in 100 → 1000 Hz
resonance frequency: 43 kHz	noise amplitude: 0.3 Å in 0.3 → 3000 Hz

Table 6.6 Magnetic force microscopy (Mamin et al. 1988).

Lever and tip specifications	Performance highlights
material: Ni	amplitude of vibration: 30 to 100 Å
shape: bent rod	resolution: 2 μm
length: 670 μm	
base diameter: 10 μm	
end diameter: 1000 Å	
constant: 0.4 N/m	
resonance frequency: 26 kHz	

Table 6.7 Electric force microscopy (Stern et al. 1988).

Lever and tip specifications	Performance highlights
material: Ni	smallest detectable
shape: bent rod	charge: 100 electrons
length: 560 μm	
base diameter: 5 μm	
constant: 0.06 N/m	
resonance frequency: 11 kHz	

6.4.3. Differential Fiber-Coupled Homodyne Detection System

Rugar et al. (1989) and Breen et al. (1990) employed the system shown schematically in Fig. 6.1 where a multimode laser diode replaces the HeNe laser. The laser-diode light is coupled into the input of a 2x2 single-mode directional coupler that directs part of the light into a photodetector. The remainder is guided by the fiber, which is placed up to 4 μm from the lever supporting the force-sensing tip. The front face of the fiber and the lever constitute a Fabry Perot with thickness ℓ, whose back reflection retraces its original path until it arrives at the directional coupler. Here part of the beam is directed into a second photodetector. The first photodetector serves as a reference signal and the second performs the homodyne detection. The current of the two photodetectors is compared electronically and the difference between these two is used to produce an image of the forces. The differential method has a large common-mode rejection, so that noise entering the two photodetectors with the same magnitude and phase cancels out. Tables 6.8 and 6.9 detail specifications and performances of the systems.

Table 6.8 Atomic force microscopy (Rugar et al. 1989).

Lever and tip specifications	Performance highlights
material: SiN	noise equivalent amplitude: 0.1 Å in 1 kHz bandwidth laser rms wavelength variation: 1.8×10^{-6} for 1 Hz - 1 kHz resolution: atomic

Table 6.9 Magnetic force microscopy (Rugar et al. 1989).

Lever and tip specifications	Performance highlights
material: Ni diameter: 2 - 4 μm	noise equivalent amplitude: 0.001 - 0.1 Å/$\sqrt{\text{Hz}}$ resolution: submicrometer

6.5. Summary

The main feature of the differential homodyne detection system is its common-mode rejection. The photocurrent generated by this system at the frequencies Ω and 2Ω are

$$i_{\Omega} = \frac{1}{2}\eta F P_i \frac{4\pi}{\lambda} A \sin(\Omega t)$$

and

$$i_{2\Omega} = \frac{1}{8}\eta F P_i \left(\frac{4\pi}{\lambda} A\right)^2 \cos(2\Omega t),$$

where η is the quantum efficiency, P_i is the power of the beam incident on the Fabry-Perot interferometer consisting of the optical flat and lever, F is a function of the reflectivity of the Fabry-Perot interferometer, λ is the optical wavelength, and A and Ω are the amplitude and frequency of the vibration of the lever.

7
Heterodyne Detection System

7.1. Introduction

In a heterodyne detection system (Martin and Wickramasinghe 1987), shown schematically in Fig. 7.1, the first beam splitter divides into two components. One passes through an acousto-optic modulator that shifts the beam frequency by Ω_m, and the other is reflected onto a mirror as a reference beam. The beam with the shifted frequency, serving as the signal beam, passes through a polarizing beam splitter, a quarter-wave plate, and finally, a microscope objective that focuses it onto the lever supporting the force-sensing tip. The lever reflects the beam back through the microscope objective and quarter-wave plate, which rotates the polarization on the two passes by 90^0. The polarizing beam splitter then deflects the beam through an analyzer that adjusts the relative power of the beam incident on the photodetector. The reference beam is deflected by a second mirror, passes through the polarizing beam splitter and analyzer, and is incident on the same photodetector. The reference and signal beams interfere on the photodetector, which generates a current consisting of a spectrum of frequencies. The photocurrent is fed into a single side-band receiver driving a phase-sensitive detector that provides the signal used to display the force acting on the tip.

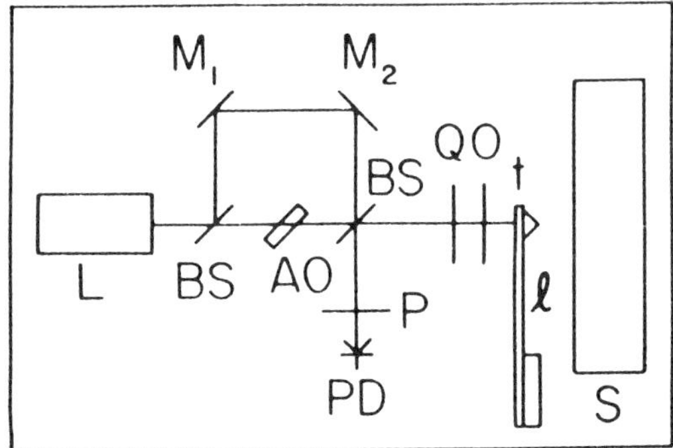

Fig. 7.1 The heterodyne detection system, showing the laser (L), the two mirrors (M_1 and M_2), beam splitter (BS), the acousto-optic modulator (AO), the quarter-wave plate (Q), the lever (ℓ) supporting the force-sensing tip (t), the sample (S), the microscope objective (O), the polarizer (P), and the photodetector (PD).

7.2. Theory

The heterodyne detection system can be analyzed by calculating the interference of the optical fields of a reference beam, E_r, and the signal beam E_s. Let the fields of these two beams be given by

$$E_r = E_{r_0} \sin(\omega t) \tag{7.1}$$

and

$$E_s = E_{s_0} \sin[\omega t + \Omega_m t + \theta_t] , \tag{7.2}$$

where ω is the optical frequency. The phase shift θ_t, given by

$$\theta_t = \frac{4\pi}{\lambda} z , \tag{7.3}$$

corresponds to the optical pathlength difference of the two arms of the heterodyne interferometer and

$$z = z_0 + A \sin(\Omega t) , \tag{7.4}$$

where A and Ω are the the amplitude and frequency of vibration of the lever. Defining θ_0 and θ by

$$\theta_0 = \frac{4\pi}{\lambda} z_0 \tag{7.5}$$

and

$$\theta = \frac{4\pi}{\lambda} A , \tag{7.6}$$

we can rewrite Eq. (7.2) as

$$E_s = E_{s_0} \sin[\omega t + \Omega_m t + \theta_0 + \theta \sin(\Omega t)] . \tag{7.7}$$

Decomposing Eq. (7.7) and retaining only the component that has the $\sin(\omega t)$ factor yields

$$E_s = E_{s_0} \sin(\omega t) \\ x[\cos(\Omega_m t + \theta_0) \cos[\theta \sin(\Omega t)] - [\sin(\Omega_m t + \theta_0) \sin[\theta \sin(\Omega t)]] . \tag{7.8}$$

The interference of the signal and reference beams on the photodetector generates a photocurrent, i, given by

$$i = \eta P_{pd} \left[\cos(\Omega_m t + \theta) \left[1 - \frac{\theta^2}{2} \sin^2(\Omega t) \right] - \theta \sin(\Omega_m t + \theta_0) \sin(\Omega t) \right] \quad (7.9)$$

Here

$$P_{pd} = 2\sqrt{P_r P_s} , \quad (7.10)$$

where P_s and P_r are the average power in the signal and reference arms. We can decompose Eq. (7.9) once and get

$$i = \eta P_{pd} [\cos(\Omega_m t + \theta_0)$$

$$- \frac{\theta^2}{2} \cos(\Omega_m t + \theta_0) \sin^2(\Omega t) - \theta \sin(\Omega_m t + \theta_0) \sin(\Omega t)] , \quad (7.11)$$

and a second time, and get

$$i = \eta P_{pd} [\cos(\Omega_m t + \theta_0) - \frac{\theta^2}{4} \cos(\Omega_m t + \theta_0)$$

$$+ \frac{\theta^2}{4} \cos(\Omega_m t + \theta_0) \cos(2\Omega t) - \theta \sin(\Omega_m t + \theta_0) \sin(\Omega t)] . \quad (7.12)$$

Using the identities

$$\sin(\alpha) \sin(\beta) = - \frac{1}{2} [\cos(\alpha + \beta) - \cos(\alpha - \beta)] \quad (7.13)$$

and

$$\cos(\alpha) \cos(\beta) = \frac{1}{2} [\cos(\alpha + \beta) + \cos(\alpha - \beta)] , \quad (7.14)$$

we find that the photocurrent i is

$$i = \eta P_{pd} \left[\cos(\Omega_m t + \theta_0) \left[1 - \frac{\theta^2}{4} \right] \right.$$

$$+ \frac{\theta^2}{8} [\cos[(\Omega_m - 2\Omega)t + \theta_0] + \cos[(\Omega_m + 2\Omega)t + \theta_0]]$$

$$\left. + \frac{\theta}{2} [\cos[(\Omega_m + \Omega)t + \theta_0] - \cos[(\Omega_m - \Omega)t + \theta_0]] \right] . \quad (7.15)$$

The photocurrent, therefore, consists of a central component oscillating at an angular frequency, Ω_m, and sidebands at $\Omega_m \pm \Omega$ and $\Omega_m \pm 2\Omega$, which are

$$i_{\Omega_m} = \eta P_{pd} \left[1 - \frac{\theta^2}{4} \right] \cos(\Omega_m t + \theta_0) , \quad (7.16)$$

$$i_{\Omega_m + \Omega} = \pm \eta P_{pd} \frac{\theta}{2} \cos[(\Omega_m \pm \Omega)t + \theta_0] , \quad (7.17)$$

and

$$i_{\Omega_m \pm 2\Omega} = \pm \eta P_{pd} \frac{\theta^2}{8} \cos[(\Omega_m \pm 2\Omega)t + \theta_0] . \quad (7.18)$$

Electronically mixing the central component i_{Ω_m} with the sideband $i_{\Omega_m + \Omega}$ gives a current

$$i = \alpha \, i_{\Omega_m} \, i_{\Omega_m + \Omega} , \quad (7.19)$$

where α is a constant determined by the receiver. Using Eq. (7.14) to decompose the two cosine functions yields

$$i = \alpha \eta^2 P_{pd}^2 \frac{2\pi}{\lambda} \left[1 - \frac{\pi^2}{\lambda^2} A^2 \right] A \, [\cos[(2\Omega_m + \Omega)t + 2\theta_0] + \cos(\Omega t)] . \quad (7.20)$$

The single sideband receiver can reject the $\cos[(2\Omega_m + \Omega)t + 2\theta_0)]$ component leaving an output current

$$i = \alpha \eta^2 P_{pd}^2 \frac{2\pi}{\lambda} A \cos(\Omega t) , \quad (7.21)$$

which is proportional to the amplitude of vibration of the lever. In contrast to homodyne detection, we find the important result that the phase angle θ_0 cancels out by the mixing of i_Ω with $i_{\Omega_m - \Omega}$, both of which contain θ_0, making heterodyne detection immune to drifts in z_0.

7.3. Noise Considerations

7.3.1. Optical Pathlength Drift

The heterodyne detection system eliminates the effects of the drift in the optical pathlength.

7.3.2. Johnson Noise

The Johnson noise generated by the detector is

$$\langle \delta i^2 \rangle_J = 4KTB/R \ , \tag{7.22}$$

and in comparison to the other noise contributions, can be neglected.

7.3.3. Laser Noise

Using Eq. (3.24), we find that the photocurrent noise of the laser is

$$\langle \delta i^2 \rangle_L = \eta^2 \, (P_r^2 + P_s^2) \ \mathrm{RIN} \ . \tag{7.23}$$

7.3.4. Lever Thermal Noise

On resonance, the lever noise translates into a current noise of the lever

$$\langle \delta i^2 \rangle_\ell = \eta^2 P_{pd}^2 \ \frac{4\pi^2}{\lambda^2} \ \frac{4KTBQ}{\omega_0 k} \ . \tag{7.24}$$

Off resonance, the current noise of the lever is

$$\langle \delta i^2 \rangle_\ell = \eta^2 P_{pd}^2 \ \frac{4\pi^2}{\lambda^2} \ \frac{4KTB}{Q\omega_0 k} \ . \tag{7.25}$$

7.3.5. Shot Noise

The shot noise generated by the detector is

$$\langle \delta i^2 \rangle_s = 2e\eta(P_r + P_s)B \ . \tag{7.26}$$

7.3.6. Signal-to-Noise Ratio

Neglecting Johnson noise, which is the smallest contribution, the signal-to-noise ratio attributable to the other contributions for on- and off-resonance operation is

$$\frac{i}{\langle \delta i^2 \rangle^{1/2}} = \frac{4\pi}{\lambda} \frac{A}{\sqrt{(p_r^2 + p_s^2)/p_{pd}^2 \ \mathrm{RIN} + \dfrac{16\pi^2}{\lambda^2} \dfrac{4KTBQ}{\omega_0 k} + \dfrac{8(P_r + P_s)eB}{\eta P_d^2}}} \tag{7.27}$$

and

$$\frac{i}{\langle \delta i^2 \rangle^{1/2}} = \frac{4\pi}{\lambda} \frac{A}{\sqrt{(p_r^2 + p_s^2)/p_{pd}^2 \ \mathrm{RIN} + \dfrac{16\pi^2}{\lambda^2} \dfrac{4KTB}{Q\omega k} + \dfrac{8(P_r + P_s)eB}{\eta P_d^2}}} . \tag{7.28}$$

7.4. Performance

Martin and Wickramasinghe (1987), Martin et al. (1987, 1988a,b), Abraham et al. (1988a,b), Hobbs et al. (1989), Sueoka et al. (1990), Bartolini et al. (1990), Nonnenmacher and Wickramasinghe (1992), and Oshio et al. (1992) employed a system similar to that illustrated in Fig. 5.1 for atomic, magnetic, and electric force microscopy. Instead of the two mirrors, they later used a Dove prism to match the optical pathlength of the two arms of the interferometer. The latest version of their electronics has been upgraded with a custom-designed lock-in amplifier to servo on the phase of the signal rather than on the amplitude. Tables 7.1 through 7.5 detail specifications and performance of various heterodyne detection systems.

Table 7.1 Atomic force microscopy (Martin et al. 1987).

Lever and tip specifications	Performance highlights
material: tungsten	amplitude of vibration: 1 nm
shape: bent rod	type of force: attractive
length: 460 μm	smallest detectable force: 3×10^{-13} N
bend length: 40 μm	smallest force derivative: 1.5×10^{-4} N/m
tip diameter: 15 μm	tip-sample distance: 3 → 18 nm
end diameter: 0.1 μm	force range: $1.5\times10^{-9} \rightarrow 10^{-11}$ N
spring constant: 7.5 N/m	force derivative range: 0.01 → 2 N/m
resonance frequency: 72 kHz	maximum force: 6.4×10^{-7} N
quality factor = 190	spatial resolution: 5 nm
	signal to noise: 1000 (B = 1, f' = 0.1 N/m)

Table 7.2 Magnetic force microscopy (Martin and Wickramasinghe 1987, Martin et al. 1988a).

Lever and tip specifications	Performance highlights
material: iron	magnetization: perpendicular
shape: rod	tip-sample distance = 20 nm → 120 nm
length: 0.6 mm (1 mm)	force range: 10^{-12} N → 35×10^{-12}
bend length: 40 μm	force derivative: $10^{-4} \rightarrow 2\times10^{-3}$ N
base diameter: 2 μm (25 μm)	spatial resolution: 100 nm
tip diameter: 0.1 μm	
spring constant: 0.08 N/m	
resonance frequency: 20 kHz	
quality factor: 370	

Table 7.3 Magnetic force microscopy (Abraham et al. 1988a,b).

Lever and tip specifications	Performance highlights
material: music wire shape: 45^0 bent rod length: 350 μm tip length: 80 μm end diameter: 2000 Å	magnetization: in-plane

Table 7.4 Magnetic force microscopy (Hobbs et al. 1989).

Lever and tip specifications	Performance highlights
material: Ni shape: bent rod length: 200, 260 μm bend length: 50, 60 μm base diameter: 50 μm tip diameter: 400, 500 Å resonance frequency: few 10 kHz quality factor =200	magnetization: in plane spatial resolution: 250 Å

Table 7.5 Electric force microscopy (Martin et al. 1988b).

Lever and tip specifications	Performance highlights
material: tungsten shape: rod spring constant: 19 N/m resonance frequency: 109 kHz	smallest force: 10^{-10} N smallest capacitance: 10^{-19} F tip-sample distance: 50 nm spatial resolution: 100 nm

7.5. Summary

The unique feature of the heterodyne detection system is that it is independent of optical pathlength drifts. The photocurrent generated by this system is

$$i = \alpha\eta^2 P_{pd}^2 \frac{2\pi}{\lambda} A \cos(\Omega t) ,$$

where α is a constant determined by the receiver, η is the quantum efficiency, P_{pd} is the power of the beam incident on the photodetector, λ is the optical wavelength, and A and Ω are the amplitude and frequency of vibration of the lever.

8
Laser–Diode Feedback Detection System

8.1. Introduction

Lasers are notorious for their sensitivity to optical feedback, and even a minute amount of light emitted by the laser that is fed back into its cavity can affect its operation drastically. Under certain optical feedback conditions, the operation of the laser can become noisy, bistable, or even chaotic. We can, however, take advantage of this sensitivity and use the laser to detect small variations in this feedback. In a laser-diode feedback detection system, shown schematically in Fig. 8.1, a lever supporting the force-sensing tip is mounted several micrometers from the front facet of the laser. The lever and front-facet combination act as a lossy Fabry Perot, whose reflectivity serves as the effective reflectivity of the front facet of the laser. The optical losses are the result of diffraction effects where successive reflections between the front facet and lever decrease for higher orders. With a round lever, for example, we can use only the first-order reflection. The system described in this chapter is based on laser-diode feedback and has unique properties that make it an attractive approach for monitoring the minute vibrations of an optical lever in a scanning force microscope. In contrast to the other optical methods, namely homodyne, heterodyne, beam deflection, and polarization, this system is

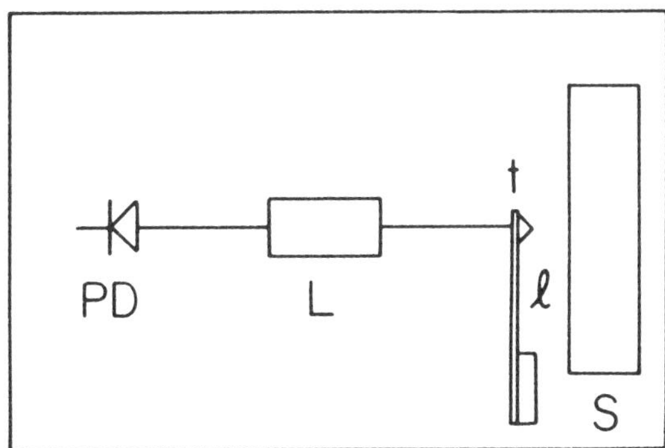

Fig. 8.1 The laser-diode feedback detection system, showing the laser (L), the lever (ℓ) supporting the force-sensing tip (t), the sample (S), and the photodetector (PD).

simple to assemble and align and has few components. The theory of operation described here (Dandridge et al. 1980b, Miles et al. 1983), however, is more complicated than that of the other systems because the optical and electronic processes occur inside the cavity of the laser that acts as a nonlinear medium.

8.2. Theory

Figure 8.2 shows a lever placed a distance z_0 away from the front facet of a laser diode acting as an external mirror with reflectivity r_m. Let the lever vibrate with an amplitude A and frequency Ω, such that the laser front-facet-to-lever distance z is given by

$$z = z_0 + A \sin(\Omega t) \ . \tag{8.1}$$

The laser cavity will be represented by the reflectivities of its back and front facets, $R_1 = |r_1|^2$ and $R_2 = |r_2|^2$, respectively, separated by the laser cavity whose length ℓ_L is assumed to be much larger than z. Multiple reflections between the front facet and the lever will give rise to an effective reflectivity, $R_e = |r_e|^2$, given by

$$r_e \exp(i\theta_e) = r_2 - \frac{1 - |r_2|^2}{r_2} \sum_{n=1}^{\infty} C_n(nz) \left[-r_2 r_m \exp \frac{4\pi i z}{\lambda} \right]^n \tag{8.2}$$

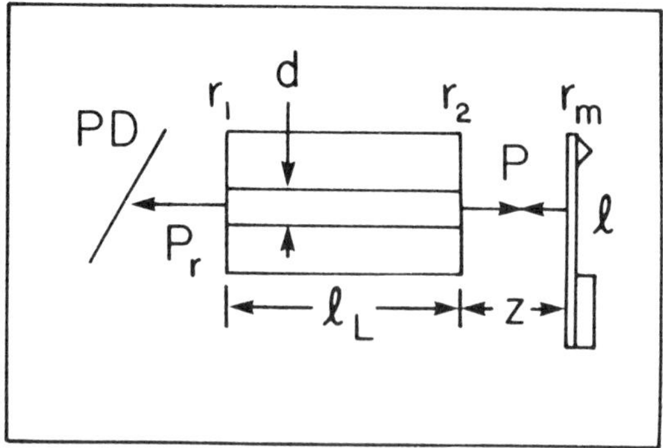

Fig. 8.2 The geometry of the laser diode where d and ℓ_L are the thickness and length of the active layer of the laser, r_1 and r_2 are the reflectivity coefficients of the rear and front facets of the laser, r_m is the reflectivity coefficient of the lever, z is the front-facet-to-lever distance, and PD is the photodetector, which is integrated into the housing of the laser.

Here $C_n(nz)$ accounts for diffraction effects arising from the geometries of the laser facet, the lever, and their distance. The diffraction attributable to the small spot size of the laser can be treated using the Fresnel numbers

$$F_a(nz) = \frac{a^2}{\lambda nz} \tag{8.3}$$

and

$$F_b(nz) = \frac{b^2}{\lambda nz}, \tag{8.4}$$

where 2a and 2b are the width and length of the emission spot on the facet of the laser. For $F_a(nz)$ and $F_b(nz) >> 1$, $F_a(nz)$ and $F_b(nz) << 1$, and $F_a(nz) < 0.8$ and $F_b(nz) > 0.8$, we can approximate the coefficients $C_n(nz)$ by

$$C_n(nz) = [1 - 0.3\,[F_a(nz)]^{-3/2} - 0.3\,[F_b(nz)]^{-3/2}]^{1/2}, \tag{8.5}$$

$$C_n(nz) = \gamma\, F_a(nz)\, F_b(nz), \tag{8.6}$$

and

$$C_n(nz) = [\gamma F_a(nz)\,(1 - 0.3[F_b(nz)]^{-3/2})]^{1/2}, \tag{8.7}$$

respectively, where $\gamma = 0.0072$. Using laser-diode rate equations, the threshold current of the laser, i_{th}, is (Miles et al. 1983)

$$i_{th}(z) = \frac{ed}{2\sigma\tau_c\eta'}\left[\gamma_i - \frac{1}{2\ell_L}\log_e[R_1R_e(z)]\right], \tag{8.8}$$

where e is the electronic charge, d is the thickness of the active layer of the laser, σ is the stimulated emission cross section for the laser transition, τ_c is the lifetime of the injected free carriers, η' is the quantum efficiency of converting current into light, and γ_i is the internal loss term. As a result of modulating the lever distance, the threshold current of the laser will be modulated through the reflectivity R_e. We can now use Eqs. (8.2) and (8.8) and find the output power of the laser from

$$P(z) = \eta'[i_d - i_{th}(z)], \tag{8.9}$$

where i_d is the driving current. The photocurrent of the photodiode monitoring the output power from the back facet of the laser, P_r, is given in terms of the power P emitted from the front facet of the laser by

$$P_r = P \frac{1 - R_1}{1 - R_e} . \tag{8.10}$$

To obtain the first harmonic of the photocurrent, we operate at the steepest slope of the curve $P(z)$, denoted here by β, where

$$\beta = \left[\frac{\partial P}{\partial z} \right]_{max} . \tag{8.11}$$

The photocurrents at dc and at Ω are, therefore,

$$i_{dc} = \eta P_r \tag{8.12}$$

and

$$i_\Omega = \eta \beta A \sin(\Omega t) . \tag{8.13}$$

We can obtain an approximate value for β by taking the variations of Eqs. (8.2), (8.8), and (8.9), obtaining

$$\delta R_e = [1 - |r_2|^2]\, C_1(z)\, r_m\, 2r_2 \frac{4\pi}{\lambda} A \sin(\Omega t) , \tag{8.14}$$

$$\delta i_{th}(z) = \left[\frac{ed}{2\sigma\tau_c\eta'} \frac{1}{2R_e \ell_L} \right] \delta R_e , \tag{8.15}$$

and

$$\delta P(z) = \eta' \, \delta i_{th}(z)] . \tag{8.16}$$

Plugging Eqs. (8.10), (8.14), and (8.15) into (8.16) yields

$$\beta = \frac{4\pi}{\lambda} \frac{1 - R_1}{1 - R_e} \eta' \frac{ed}{\sigma\tau_c\eta'} \frac{r_2}{2R_e \ell_L} (1 - |r_2|^2)\, C_1\, r_m , \tag{8.17}$$

which is a constant as long as z_0 is maintained at a constant value. We find, therefore, that to the first order the photocurrent i_Ω is linear in

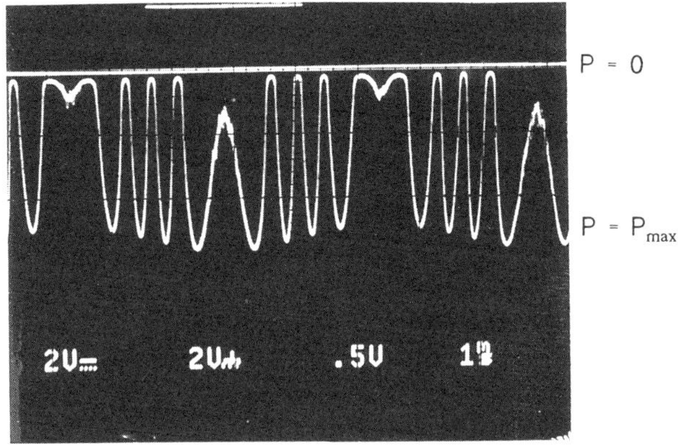

Fig. 8.3 The visibility of fringes of the laser-diode feedback system.

the deflection of the lever and can be used to monitor its amplitude of vibration. Figure 8.3 shows the power output of the laser diode as a function of the lever to front facet of the laser distance. Note that moving the lever by $\lambda/4 \simeq 2000$ Å changes the optical power between 0 and $\simeq 3$ mW.

8.3. Noise Considerations

Optical feedback modifies the threshold current, operating frequency, and emission linewidth of a laser diode and suppresses its sidemodes. It is not possible, therefore, to present the noise figures in a simple analytic form. We can, however, simplify noise considerations by using the parameters P_r and β, which can be obtained experimentally.

8.3.1. Optical Pathlength Drift

The laser-diode feedback detection system suffers from effects of drift in the optical pathlength z_0 that result from both ambient temperature changes and the effects of heat generated in the laser diode. These can be minimized by choosing z_0 to be as small as several micrometers.

8.3.2. Johnson Noise

The Johnson noise generated by the detector is

$$\langle \delta i^2 \rangle_J = 4KTB/R \ , \tag{8.18}$$

and in comparison to the other noise contributions, can be neglected.

8.3.3. Laser Noise

The current noise of the laser is

$$\langle \delta i^2 \rangle_L = \eta^2 P^2{}_r \ \mathrm{RIN} \, . \tag{8.19}$$

8.3.4. Lever Thermal Noise

On resonance, the lever noise translates into a current noise of the lever

$$\langle \delta i^2 \rangle_\ell = \beta^2 \eta^2 \, \frac{4KTBQ}{\omega_0 k} \, . \tag{8.20}$$

Off resonance, we get

$$\langle \delta i^2 \rangle_\ell = \beta^2 \eta^2 \, \frac{4KTB}{Q\omega_0 k} \tag{8.21}$$

for the current noise of the lever.

8.3.5. Shot Noise

The shot noise generated by the detector is

$$\langle \delta i^2 \rangle_s = 2e\eta P_r B \, . \tag{8.22}$$

8.3.6. Signal-to-Noise Ratio

Neglecting the Johnson noise, which is the smallest contribution, the signal-to-noise ratio attributable to all noise contributions for on- and off-resonance operation is, respectively,

$$\frac{i}{\langle \delta i^2 \rangle^{1/2}} = \frac{\beta}{P_r} \, \frac{A}{\sqrt{\mathrm{RIN} + \dfrac{\beta^2}{P^2{}_r} \, \dfrac{4KTBQ}{\omega_0 k} + \dfrac{2eB}{\eta P_r}}} \tag{8.23}$$

and

$$\frac{i}{\langle \delta i^2 \rangle^{1/2}} = \frac{\beta}{P_r} \, \frac{A}{\sqrt{\mathrm{RIN} + \dfrac{\beta^2}{P^2{}_r} \, \dfrac{4KTB}{Q\omega_0 k} + \dfrac{2eB}{\eta P_r}}} \, . \tag{8.24}$$

8.4. Performance

Sarid et al. (1988, 1989a, 1989b, 1992a), Sarid and Elings (1991), and Denk and Pohl (1991) employed the system shown in Fig. 8.1. Here a Hitachi LT022MDO, 790-nm, 3-mW laser diode replaces all the optical components used in the scanning force microscope, and a photodetector integrated into the housing of the laser diode monitors the optical power of the laser light emitted from its rear facet. The lever is mounted on a bimorph and is placed several micrometers from the front facet of the laser, forming a compact integrated force-sensing head. The light emitted from the front facet is incident on a lever supporting the force-sensing tip, and its reflection is coupled back into the laser cavity. The original system, using three piezoelectric-driven translators, served as magnetic and electric force microscopes. The laser-diode integrated head, in its new implementation, is mounted into a removable or stand-alone head of a commercial STM (Digital Instruments, Nanoscope II and III). Tables 8.1 through 8.3 detail specifications and performance of the laser-diode feedback system.

Table 8.1 Magnetic force microscopy (Sarid et al. 1988).

Lever and tip specifications	Performance highlights
material: nickel	amplitude of vibration: 10 nm
shape: bent rod	spatial resolution: 800 Å
length: 1 mm	signal to noise: 0.003 Å/ $Hz^{1/2}$ at 1 kHz
bend length: ≃ 200 μm	
base diameter: 25 μm	
end diameter: 0.1 μm	
spring constant: 1 N/m	
resonance frequency: 2 kHz	
quality factor = 100	

Table 8.2 Electric force microscopy (Sarid et al. 1989a, 1990).

Lever and tip specifications	Performance highlights
material: nickel shape: bent rod length: 1 mm bend length: $\simeq$ 200 μm base diameter: 25 μm end diameter: 0.1 μm constant: 1 N/m resonance frequency: 2 kHz quality factor = 100	amplitude of vibration: 10 nm spatial resolution: 1000 Å signal to noise: 0.003 Å / $Hz^{1/2}$ at 1 kHz

Table 8.3 Atomic force microscopy (Sarid et al. 1992a).

Lever and tip specifications	Performance highlights
material: Si_3N_4 shape: triangle	resolution: atomic

8.5. Summary

The laser-diode feedback detection system is unique because of its compactness and because it requires no optical elements. The photocurrent generated by this system is given by

$$i = \eta\beta A \sin(\Omega t) ,$$

where β is a constant and A and Ω are the amplitude and frequency of the vibration of the lever.

9
Polarization Detection System

9.1. Introduction

The polarization detection system (Schönenberger and Alvarado 1990a and Anselmetti et al. 1992), shown schematically in Fig. 9.1, which is similar to that of den Boef (1989, 1990, 1991), differs from the other optical detection systems in that the amplitude of vibration of the lever is first converted into a polarization modulation that is subsequently converted into amplitude modulation. The conversion from polarization to amplitude modulation is carried out by two polarizing prisms rotated 45^0 relative to each other that produce an interference between s- and p-polarized fields. The output from the differential system, which has a common-mode rejection that cancels most of the laser noise, is used to image the forces across the sample. (Hane et al. 1990 show a system using two gratings to direct the light on both ends of the lever.)

9.2. Theory

It is convenient to analyze the differential polarization detection system by following the path of the polarized laser beam through the optical elements on its way to the lever supporting the force sensing tip and back into the two photodetectors. The beam first passes through a Faraday isolator that protects the laser from back reflections and emerges as a plane-polarized beam whose plane of polarization is 45^0 in respect to the calcite prism. The beam can be decomposed into two components having mutually perpendicular polarizations with field and power denoted by E_s, P_s, E_r, and P_r, respectively. These two beams, which share a common trajectory, pass through a Soleil Babinet compensator that introduces an adjustable phase shift, ψ, between them. After being expanded, they pass through a beam splitter, a microscope objective, and a calcite prism. The calcite prism separates the beams spatially, focusing the reference beam on the base of the vibrating lever and the signal beam on the flexible part of the lever close to the force-sensing tip. Let us denote the optical pathlength of the reference and signal beams, between the calcite prism and the lever, by z_0 and z, respectively, related by

$$z = z_0 + A\sin(\Omega t) , \tag{9.1}$$

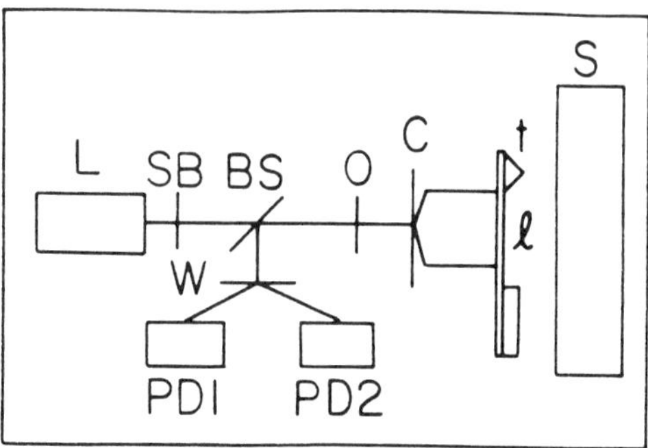

Fig. 9.1 The polarization detection system, showing the laser (L), the Soleil Babinet compensator (SB), the beam splitter (BS), the lever (ℓ) supporting the force-sensing tip (t), the sample (S), the microscope objective (O), the calcite prism (C), the Wollaston prism (W), and the photodetectors (PD_1 and PD_2).

where A and Ω are the amplitude and frequency of the vibration of the lever. The phase angle acquired by the reference beam during its round trip is

$$\theta_0 = \frac{4\pi}{\lambda} z_0 , \tag{9.2}$$

while that of the signal beam is

$$\theta = \theta_0 + \frac{4\pi}{\lambda} A \sin(\Omega t) . \tag{9.3}$$

The signal beam, therefore, undergoes an extra phase modulation, $\delta\theta$, on reflection from the lever, given by

$$\delta\theta = \frac{4\pi}{\lambda} A \sin(\Omega t) . \tag{9.4}$$

The reference and signal beams, on their way back from the lever, are recombined into a single beam by the calcite prism. From there they continue toward the beam splitter, which deflects them onto a Wollaston prism rotated 45° relative to the calcite prism. Because the two beams passed twice through the beam splitter, their field intensity and power are reduced to $E_s/2$, $P_s/4$, $E_p/2$, and $P_r/4$, respectively. Figure 9.2 shows the E_s and E_p fields that interfere inside the Wollaston prism. The interference generates the E_1 and E_2 fields, which are given by

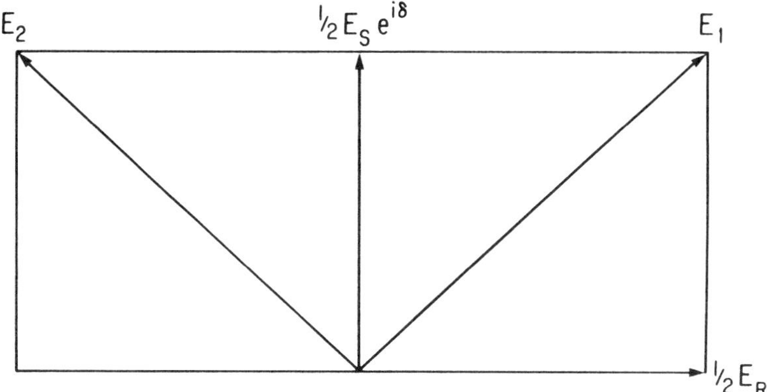

Fig. 9.2 The interference of fields E_s and E_p generates fields E_1 and E_2 inside the Wollaston prism.

$$|E_1|^2 = \frac{1}{8}\,|E_s + E_p \exp(\psi + \delta\theta)|^2 \tag{9.5}$$

and

$$|E_2|^2 = \frac{1}{8}\,|-E_s + E_p \exp(\psi + \delta\theta)|^2 \,. \tag{9.6}$$

Assuming again that the two beams are of equal power, $P_s = P_r = P/2$, we can write the power $P_1 = |E_1|^2$ and $P_2 = |E_2|^2$ as

$$P_1 = \frac{1}{8}\,P\,[1 + \cos(\psi + \delta\theta)] \tag{9.7}$$

and

$$P_2 = \frac{1}{8}\,P\,[1 - \cos(\psi + \delta\theta)] \,. \tag{9.8}$$

The currents in the two photodetectors will, therefore, be

$$i_1 = \frac{1}{8}\,\eta P\,[1 + \cos(\psi + \delta\theta)] \tag{9.9}$$

and

$$i_2 = \frac{1}{8} \eta P \, [1 - \cos(\psi + \delta\theta)] \; , \tag{9.10}$$

where η is the quantum efficiency of the photodetectors. The case where the two beams do not have the same power can be analyzed by defining the visibility of fringes, V, by

$$V = 2 \, \frac{\sqrt{P_r P_s}}{P_r + P_s} \; , \tag{9.11}$$

and get for the ratio of the difference and sum of the two photocurrents, β,

$$\beta = \frac{i_1 - i_2}{i_1 + i_2} = V \cos(\psi + \delta\theta) \; . \tag{9.12}$$

The phase shift ψ generated by the Soleil Babinet compensator can now be scanned so that β goes between its minimum and maximum values which, ideally, are unity and zero. During an experiment, however, since we want to measure small values of $\delta\theta$, we will adjust the phase angle ψ to equal $\pm\pi/2$, obtaining

$$\cos(\psi + \delta\theta) = \sin(\delta\theta) \simeq \delta\theta \; . \tag{9.13}$$

We note that the value of V is fixed during an experiment because it depends on the power of the two beams and not on their relative phase shift. The value of β, given by

$$\beta = V \, \frac{4\pi}{\lambda} \, \mathrm{A}\sin(\Omega t) \; , \tag{9.14}$$

is the signal that is used by the computer to generate the image of the forces across the surface of the sample.

9.3. Noise Considerations

To calculate the signal and the noise of the system, we consider the photocurrents at the input of the differential amplifier,

$$i_1 = \frac{1}{8}\eta P\left[1 + \frac{4\pi}{\lambda}\mathrm{Asin}(\Omega)\right] \tag{9.15}$$

and

$$i_2 = \frac{1}{8}\eta P\left[1 - \frac{4\pi}{\lambda}\mathrm{Asin}(\theta)\right], \tag{9.16}$$

where the signal is

$$i = i_1 - i_2 = \frac{1}{4}\eta P\frac{4\pi}{\lambda}\mathrm{Asin}(\Omega). \tag{9.17}$$

9.3.1. Optical Pathlength Drift

The thermal drift of the polarization detection system can be minimized by making the optical pathlength difference between the reference and signal beams small.

9.3.2. Johnson Noise

The Johnson noise generated by each photodetector is

$$\langle \delta i_1{}^2 \rangle_J = 4KTB/R \tag{9.18}$$

and

$$\langle \delta i_2{}^2 \rangle_J = 4KTB/R, \tag{9.19}$$

and the noise at the output of the differential amplifier is

$$\langle \delta i^2 \rangle_J = 8KTB/R, \tag{9.20}$$

which is larger by a factor of 2 than the noise of each detector. In comparison to the other noise contributions, it can be neglected.

9.3.3. Laser Noise

The laser noise in each detector is

$$\langle \delta i^2{}_1 \rangle_L = \frac{1}{64} \eta^2 \text{ RIN } [P + P\delta\theta]^2 \tag{9.21}$$

and

$$\langle \delta i^2{}_2 \rangle_L = \frac{1}{64} \eta^2 \text{ RIN } [P - P\delta\theta]^2 . \tag{9.22}$$

The low-frequency noise at the output of the differential amplifier is given by subtracting Eq. (9.21) from Eq. (9.22), yielding

$$\langle \delta i^2 \rangle_L = \frac{1}{16} \text{ RIN } P^2 \, \delta\theta . \tag{9.23}$$

Since $\delta\theta << 1$, we find that the low-frequency components of the noise are reduced considerably, giving a good common-mode rejection.

9.3.4. Lever Thermal Noise

On resonance, the lever noise translates into noise in each photodetector,

$$\langle \delta i^2{}_1 \rangle_\ell = \frac{\pi^2}{2\lambda^2} \eta^2 P^2 \langle \delta A^2 \rangle \tag{9.24}$$

and

$$\langle \delta i^2{}_2 \rangle_\ell = \frac{\pi^2}{2\lambda^2} \eta^2 P^2 \langle \delta A^2 \rangle . \tag{9.25}$$

The thermal noise of the lever, at the output of the differential amplifier, will be

$$\langle \delta i^2 \rangle_\ell = \frac{\pi^2}{\lambda^2} \eta^2 P^2 \, \frac{4KTBQ}{\omega_0 k} , \tag{9.26}$$

which is larger by a factor of 2 than the noise of each detector. Off resonance, we get

$$\langle \delta i^2 \rangle_\ell = \frac{\pi^2}{\lambda^2} \eta^2 P^2 \, \frac{4KTB}{Q\omega_0 k} \tag{9.27}$$

for the noise.

9.3.5. Shot Noise

The shot noise generated by each photodetector is

$$\langle \delta i_1{}^2 \rangle_s = \frac{1}{4} e\eta PB \tag{9.28}$$

and

$$\langle \delta i_2{}^2 \rangle_s = \frac{1}{4} e\eta PB \ , \tag{9.29}$$

and the noise at the output of the differential amplifier will be

$$\langle \delta i^2 \rangle_s = \frac{1}{2} e\eta PB \ , \tag{9.30}$$

which is larger by a factor of 2 than the noise of each detector.

9.3.6. Signal-to-Noise Ratio

Neglecting Johnson and laser noise, which are the smallest contributions, the signal-to-noise ratio attributable to the other noise sources for on- and off-resonance operation, assuming that $P_r = P_s = P/2$, is

$$\frac{i}{\langle \delta i^2 \rangle^{1/2}} = \frac{\pi}{\lambda} \frac{A}{\sqrt{\frac{\pi^2}{\lambda^2} \frac{4KTBQ}{\omega_0 k} + \frac{eB}{2\eta P}}} \tag{9.31}$$

and

$$\frac{i}{\langle \delta i^2 \rangle^{1/2}} = \frac{\pi}{\lambda} \frac{A}{\sqrt{\frac{\pi^2}{\lambda^2} \frac{4KTB}{Q\omega_0 k} + \frac{eB}{2\eta P}}} \ , \tag{9.32}$$

where we assumed a visibility, $V = 1$.

9.4. Performance

Schönenberger and Alvarado (1989, 1990*a*) and Schönenberger et al. (1990) employed the system shown schematically in Fig. 9.1 for atomic and magnetic force microscopy. They found that the only contributions to the noise are lever vibration and shot noise, which were similar in magnitude. Tables 9.1 and 9.2 detail the polarization detection systems and their performance.

Table 9.1 Atomic force microscopy (Schönenberger and Alvarado 1989).

Lever and tip specifications	Performance highlights
lever material: GaAs wafer	amplitude of vibration: 0.2 Å
shape: rectangular	force resolution: 10^{-10} N
length: 3 mm	operating force derivative: 0.1 N/m
width: 50 μm	noise: 6×10^{-5} Å/$\sqrt{\text{Hz}}$ at above 1 kHz
tip material: tungsten	noise: 0.01 Å in bandwidth of 1 → 20 kHz
base diameter: 25 μm	resolution: submicrometer
length: 100 μm	noise: 0.4 Å in bandwidth of 0.01 → 1 kHz
spring constant: 20 N/m	drift: 0.01 - 0.02 Å/s
resonance frequency: 45 kHz	
quality factor: 200	

Table 9.2 Atomic and magnetic force microscopy (Schönenberger et al. 1990).

Lever and tip specification	Performance highlights
Spring constant: 1 and 5 N/m	smallest detectable force: 10^{-11} N
	smallest lever deflection: 0.05 Å at 1 Hz
	resolution: submicrometer

9.5. Summary

The main feature of the polarization detection system is its common-mode rejection. The photocurrent generated by this system is

$$i = \eta\sqrt{P_r P_s}\,\frac{8\pi}{\lambda}\,A\sin(\Omega)\ ,$$

where η is the quantum efficiency, P_r and P_s are the power of the reference and signal beams, λ is the optical wavelength, and A and Ω are the amplitude and frequency of the vibration of the lever.

10
Deflection Detection System

10.1. Introduction

In a deflection detection system (Meyer and Amer 1988a,b, 1990a, Alexander et al. 1989, Marti et al. 1990, and Hipp et al. 1992), shown in Fig. 10.1, a collimated laser beam is focused on a lever supporting the force-sensing tip, and is reflected back into two closely spaced photodetectors whose photocurrents are fed into a differential amplifier. A minute deflection of the lever causes one photodetector to collect more light than the other, and the output of the differential amplifier, which is proportional to the deflection of the lever, is used to image the forces across a sample. The system differs from the other optical systems in that all the optical elements are at a large distance from the lever, and are therefore protected from mechanical damage if the tip "crashes." Later we will describe in detail three different implementations of the deflection method. Here we analyze a deflection detection system using a simplified case, where the optical beam has a "rectangular" cross section. Although the more accurate theory, using Gaussian beams, would be somewhat more involved, we expect the result to be similar.

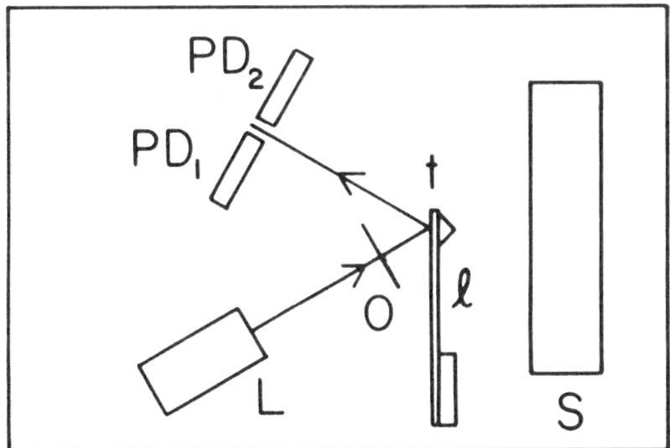

Fig. 10.1 A schematic diagram of the deflection detection system showing the laser (L), the microscope objective (O), the lever (ℓ) supporting the force-sensing tip (t), the sample (S), and the photodetectors (PD_1 and PD_2).

10.2. Theory

Consider the system shown in Fig. 10.2 where ℓ is the length of the lever supporting the force-sensing tip, A is the deflection of the lever, s is the lever-photodetector distance, d^2 is the area of the "square" profile of the optical beam incident on the plane of the two photodetectors, P is the optical power of the beam, and R and η are the load resistance and the quantum efficiency of each photodetector, respectively. Let us calculate the angle at the end of a lever in the presence of a force F acting at this point. By taking the derivative of Eq. (1.18) we get

$$\theta = \frac{F\ell^2}{2EI} . \tag{10.1}$$

Since $EI = k\ell^2/3$ and $F = kz$, for θ we get

$$\theta = \frac{3}{2} \frac{z}{\ell} . \tag{10.2}$$

A deflection of the lever by $z = A\sin(\Omega t)$, where A is the amplitude and Ω the frequency of modulation, will produce a linear deflection of the optical beam at the plane of the photodetectors, δd, given by

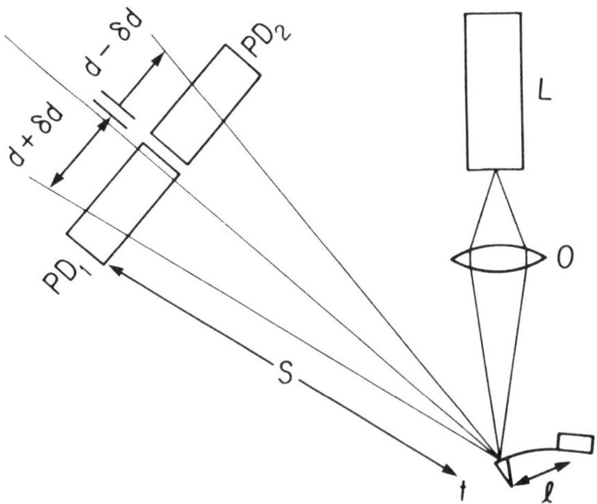

Fig. 10.2 The trajectory of the laser beam that is focused by the objective (O) on the lever (ℓ) and is reflected onto two photodetectors (PD_1 and PD_2). The length of the lever is ℓ and the lever-photodetector distance is s.

$$\delta d = 3\,\frac{s}{\ell}\,A\sin(\Omega t)\,, \tag{10.3}$$

and the optical power incident on each photodetector is

$$P_1 = \frac{1}{2}\,P\left[1 + \frac{\delta d}{d}\right] \tag{10.4}$$

and

$$P_2 = \frac{1}{2}\,P\left[1 - \frac{\delta d}{d}\right], \tag{10.5}$$

respectively. The deflection of the lever will therefore give rise to an imbalance in the power incident on the two photodetectors given by

$$\delta P = P\,\frac{\delta d}{d} = P\,\frac{3s}{\ell d}\,A\sin(\Omega t)\,. \tag{10.6}$$

The photocurrents generated by the photodetectors and their difference will then be

$$i_1 = \frac{1}{2}\,\eta P\left[1 + 3\,\frac{s}{\ell d}\,A\sin(\Omega t)\right], \tag{10.7}$$

$$i_2 = \frac{1}{2}\,\eta P\left[1 - 3\,\frac{s}{\ell d}\,A\sin(\Omega t)\right], \tag{10.8}$$

and

$$\delta i = i_1 - i_2 = \eta P\,\frac{3s}{\ell d}\,A\sin(\Omega t)\,. \tag{10.9}$$

The amplification factor β given by

$$\beta = \frac{3s}{\ell} \tag{10.10}$$

can be as large as 350, exceeding the amplification of the sensitivity of a vibrating lever due to the quality factor Q, which is usually less than 200 in air (for a discussion of the theory, see Gould et al. 1990b and Putman et al. 1992).

For a beam spot size on the lever given by $2a \times 2a$, the far-field diffraction-limited spot size on each photodetector is

$$d = \frac{1}{\pi} \frac{s}{a} . \qquad (10.11)$$

Using Eq. (10.11) in Eq. (10.9) gives

$$\delta i = 3 \frac{\pi a}{\lambda \ell} \eta \, \mathrm{PA} \sin(\Omega t) , \qquad (10.12)$$

which is independent of the lever-photodetector distances. Two-dimensional deflection methods have been discussed by Marti et al. (1990).

10.3. Noise Considerations

10.3.1. Optical Pathlength Drift

The deflection detection system in its differential implementation suffers from thermal and mechanical drift in the direction perpendicular to the optical beam, but is independent of optical pathlength drift.

10.3.2. Johnson Noise

The Johnson noise generated by each photodetector is

$$\langle \delta i_1{}^2 \rangle_J = 4KTB/R \qquad (10.13)$$

and

$$\langle \delta i_2{}^2 \rangle_J = 4KTB/R , \qquad (10.14)$$

and the noise fed into the differential amplifier becomes

$$\langle \delta i^2 \rangle_J = 8KTB/R , \qquad (10.15)$$

which is larger by a factor of 2 than that of each photodetector. In comparison with the other noise contributions, however, it can be neglected.

10.3.3. Laser Noise

The photocurrent noise in each photodetector is

$$\langle i_1^2 \rangle_L = \frac{1}{4} \eta^2 \text{ RIN } P^2 \left[1 + \frac{\delta d}{d} \right]^2 \tag{10.16}$$

and

$$\langle i_2^2 \rangle_L = \frac{1}{4} \eta^2 \text{ RIN } P^2 \left[1 - \frac{\delta d}{d} \right]^2 . \tag{10.17}$$

The low-frequency noise at the output of the differential amplifier is obtained by subtracting Eq. (10.12) from Eq. (10.13), yielding

$$\langle i^2 \rangle_L = \eta^2 \text{ RIN } P^2 \frac{\delta d}{d} . \tag{10.18}$$

Since $\delta d << d$, we find that the low-frequency components of the noise at the output of the differential amplifier are reduced considerably, giving a good common-mode rejection.

10.3.4. Lever Thermal Noise

The thermal noise of the lever translates into photocurrent noise in each photodetector

$$\langle \delta i^2{}_1 \rangle_\ell = \eta^2 P^2 \frac{9s^2}{4\ell^2 d^2} \frac{4KTBQ}{\omega_0 k} \tag{10.19}$$

and

$$\langle \delta i^2{}_2 \rangle_\ell = \eta^2 P^2 \frac{9s^2}{4\ell^2 d^2} \frac{4KTBQ}{\omega_0 k} . \tag{10.20}$$

The thermal noise of the lever, at the output of the difference amplifier, will be

$$\langle \delta i^2 \rangle_\ell = \eta^2 P \frac{9s^2}{4\ell^2 d^2} \frac{8KTBQ}{\omega_0 k} , \tag{10.21}$$

which is larger by a factor of 2 than that of each photodetector. Off resonance, we get

$$\langle \delta i^2 \rangle_\ell = \eta^2 P^2 \frac{9s^2}{4\ell^2 d^2} \frac{8KTB}{Q\omega_0 k} \tag{10.22}$$

for the noise, which is larger by a factor of 2 than that of each photodetector.

10.3.5. Shot Noise

The shot noise generated by each photodetector is

$$\langle \delta i_1{}^2 \rangle_s = 2e\eta B \frac{P}{2} \tag{10.23}$$

and

$$\langle \delta i_2{}^2 \rangle_s = 2e\eta B \frac{P}{2} , \tag{10.24}$$

and the noise at the output of the differential amplifier will be

$$\langle \delta i^2 \rangle_s = 2e\eta B\ P , \tag{10.25}$$

which is larger by a factor of 2 than that of each photodetector.

10.3.6. Signal-to-Noise Ratio

Neglecting Johnson and laser noise, which are the smallest contributions, the signal-to-noise ratio attributable to the other noise sources for on- and off-resonance operation is

$$\frac{i}{\langle \delta i^2 \rangle^{1/2}} = \frac{A}{\sqrt{\dfrac{2KTBQ}{\omega_0 k} + \dfrac{9\ell^2 d^2}{8s^2} \dfrac{eB}{\eta P}}} \tag{10.26}$$

and

$$\frac{i}{\langle \delta i^2 \rangle^{1/2}} = \frac{A}{\sqrt{\dfrac{2KTB}{Q\omega_0 k} + \dfrac{9\ell^2 d^2}{8s^2} \dfrac{eB}{\eta P}}} . \tag{10.27}$$

10.4. Performance

10.4.1. Atomic Force Microscopy (Attractive)

Meyer et al. (1988) employed the system shown schematically in Fig. 10.1. The instrument, which was operated in a UHV chamber, was a modified pocket-size STM using a tungsten stylus-lever system, chemically etched and bent. The lever was vibrated close to its resonance frequency, and the change in amplitude of vibration induced by the van der Waals attractive force derivative was measured. Table 10.1 details the lever and tip specifications and performance of this deflection detection system.

10.4.2. Atomic Force Microscopy (Repulsive)

Alexander et al. (1989) and Drake et al. (1989) employed a system similar to the one shown in Fig. 10.1. In their first version, the V-shaped lever had a diamond chip glued at its end. A shattered piece of aluminized glass served as the mirror, which also was glued to the lever. The sample was mounted on a piezo tube and the light deflected from the mirror was incident on a position-sensitive photodetector. The whole system was controlled by Nanoscope II electronics. The authors also used a lever made of a 28-μm-thick piezoelectric foil that was vibrated close to its resonance frequency. The approach of the tip to the sample was monitored by oscillating the tip and observing the amplitude of vibration. At a distance of 100 Å, the amplitude starts changing because of the interaction of the tip with van der Waals forces at the surface of the sample. Another implementation of their system uses a microfabricated lever, where the tip and sample were immersed in a liquid. Its main advantage is that the levers could be made smaller

Table 10.1 Atomic force microscopy (Meyer and Amer 1988a,b).

Lever and tip specifications	Performance highlights
length: 1 mm	thermal lever vibration: 4×10^{-3} Å/$\sqrt{\text{Hz}}$
diameter: 75 μm	shot noise: 4×10^{-4} Å/$\sqrt{\text{Hz}}$
spring constant: 10^3 N/m	force derivatives: 0.013 to 0.68 N/m
resonance frequency: 20 kHz	resolution: 240 Å
quality factor Q: 800	
mirror dimensions: 300×300 μm^2	

by using microfabrication techniques. As a result, they obtained a higher optical advantage and a gentler interaction with the sample, which was not pushed around or damaged. An advantage of operating under a liquid is that it eliminates the contamination layer that covers the tip and the sample, which can give rise to adhesion forces on the order of 10^{-7} N, large enough to make imaging with atomic resolution difficult. Tables 10.2 and 10.3 detail lever and tip specifications and performance of this system.

Table 10.2 Atomic force microscopy (Alexander et al. 1989).

Lever and tip specifications	Performance highlights
length: 1.4 mm	optical advantage: 70
diameter: 20 μm	mirror-photodetector distance: 10 cm
spring constant: 10 N/m	thermal and piezoelectric drift: 0.1 Å/sec
resonance frequency: 6 kHz	typical detected forces: 10^{-6} N
diamond tip size: 0.1×0.1 mm	typical change in forces: 4×10^{-10} N
mirror thickness: 15 μm	typical lever deflection: 0.8 Å
mirror size: 0.2×0.2 mm^2	

Table 10.3 Atomic force microscopy (Drake et al. 1989).

Lever and tip specifications	Performance highlights
length: 100 μm	optical advantage: 800
diamond tip mass: 2×10^{-10} Kg	mirror-photodetector distance: 4 cm
	detectable force: 2×10^{-9} *N*
	effective noise: 0.2 Å rms (0.1 Hz - 20 KHz)
	resolution: atomic
	resolution under water: 5 Å

10.4.3. Atomic Force Microscopy (Taubenblatt 1989)

The system here is a modified scanning tunneling microscope with a tip that could vibrate laterally parallel to the surface of a sample, close to its third-harmonic resonance frequency. Lateral forces caused by friction changed the resonance frequency which, in turn, changed the amplitude of vibration. The vibration of the tip was monitored by a laser beam passing through a beam splitter and into a reference photodetector. The remainder of the beam continued through a polarizing beam splitter and a quarter-wave plate. From there the beam was focused by a microscope objective through the back of the sample onto the tip of the STM. The beam was then reflected back, traversing the same path as before. Because it passed twice through the quarter-wave plate, its polarization was rotated by 90^0 and the polarizing beam splitter deflected it into the second photodetector. The output of the differential amplifier was used to image the lateral forces across the surface of a sample. Table 10.4 details the specifications and performance of this system.

Table 10.4 Atomic force microscopy (Taubenblatt 1989).

Lever and tip specifications	Performance highlights
material: Pt	vibration amplitude: 20 Å
length: 5.2 mm	resolution: 20 Å
diameter: 25 μm	
resonance frequency: 3.4 kHz	
quality factor: 300	

10.5. Summary

The unique features of the deflection detection system are that it is a remote sensor, it is wavelength independent, it has a large magnification, and it is differential. The photocurrent generated by this system is

$$i = \eta P \frac{2s}{\ell d} A \sin(\Omega t) ,$$

where η is the quantum efficiency in mA/mW, P is the laser power, s is the lever-photodetector distance, ℓ is the length of the lever, $2d$ is the width of the laser beam, and A and Ω are the amplitude and frequency of vibration of the lever.

11
Electric Force Microscopy

11.1. Introduction

This chapter describes the principles of operation of electric force microscopy, which maps electrostatic forces across the surface of a sample. These forces, which are always attractive, act between a conducting tip supported by a lever and a conducting sample. We start with the basic theory of electrostatics and give several examples that are representative of relevant cases. These examples include a plane-parallel capacitor, a sphere and a plane, a line charge, a uniform surface charge distributed in a strip, a charged circular disk, and strips with alternate potentials. Next we develop the theory of operation of electric force microscopy for systems using a bimorph-driven lever, a sample-driven lever, and a voltage-driven lever. We derive general expressions that give the change in amplitude of vibration of the lever in terms of the parameters of the system and discuss the signal-to-noise ratio. We then present applications and experimental results of electric force microscopy related to topography, effective dielectric constant of a thin film, deposited charges, and potentials across a sample. We conclude with a summary that gives working equations for each practical case.

11.2. Basic Concepts

We concern ourselves in this section with a series of topics in electrostatics, namely the measurement of static or slowly varying electric fields between two conductors separated by a distance z that can be partially filled with a dielectric. The treatment is based on a macroscopic approach, although electric force microscopy can detect quantities small enough to be considered microscopic, such as a small number of electrons deposited on a dielectric. We start by presenting a brief overview of the main definitions and concepts associated with problems in electrostatics. First, we define the gradient of a scalar function, $\nabla\Phi$,

$$\nabla\Phi = \frac{\partial\Phi}{\partial x}\mathbf{i}_1 + \frac{\partial\Phi}{\partial y}\mathbf{i}_2 + \frac{\partial\Phi}{\partial z}\mathbf{i}_3 , \tag{11.1}$$

the divergence of a vector field, $\nabla\cdot\mathbf{E}$,

$$\nabla \cdot \mathbf{E} = \frac{\partial E_1}{\partial x} + \frac{\partial E_2}{\partial y} + \frac{\partial E_3}{\partial z} , \qquad (11.2)$$

and the curl of a vector field, $\nabla \times \mathbf{E}$,

$$\nabla \times \mathbf{E} = \left[\frac{\partial E_3}{\partial y} - \frac{\partial E_2}{\partial z} \right] \mathbf{i}_1 - \left[\frac{\partial E_3}{\partial x} - \frac{\partial E_1}{\partial z} \right] \mathbf{i}_2 + \left[\frac{\partial E_2}{\partial x} - \frac{\partial E_1}{\partial y} \right] \mathbf{i}_3 , \qquad (11.3)$$

where $\mathbf{i}_i$ denotes unit vectors in the x, y, and z directions. Using these definitions, we write the two relevant Maxwell equations for electrostatics,

$$\nabla \cdot \mathbf{E} = \rho / \epsilon_0 \qquad (11.4)$$

and

$$\nabla \times \mathbf{E} = 0 , \qquad (11.5)$$

where ρ is the volume charge density in units of coulomb/m^3, ϵ_0 = 8.85×10^{-12} farad/m is the dielectric constant of free space, and $\mathbf{E}$ is the electric field in units of volt/m. We find that $\mathbf{E}$ has to be a gradient of a scalar potential V, in units of volt/m,

$$\mathbf{E} = -\nabla V . \qquad (11.6)$$

The electric field is normal to the surface, and the potential is constant across the surface. A useful tool for solving problems in electrostatics by invoking boundary conditions is Poisson's equation

$$\nabla^2 V = -\rho / \epsilon_0 . \qquad (11.7)$$

The force acting between two point charges, q_i, is given by Coulomb's law

$$F_{12} = \frac{1}{4\pi\epsilon_0} \frac{q_1 q_2}{r^2} \mathbf{r}_{12} , \qquad (11.8)$$

where r is the distance between the two charges and $\mathbf{r}_{12}$ is a unit vector pointing from one charge to the other. The principle of superposition, applied to a collection of charges, gives the total field as the sum of the fields resulting from the individual charges. The capacitance C, in units of farads, is defined as the ratio of the charge of a conducting structure divided by its potential

$$C = \frac{q}{V}, \tag{11.9}$$

and the energy associated with this capacitance is

$$W = \frac{1}{2}\int_{v} V\rho dv , \tag{11.10}$$

where v is the volume. The energy, force per unit area, and force derivative are given, respectively, by

$$W = \frac{1}{2}\epsilon_0 \int_{v} E^2 \, dv , \tag{11.11}$$

$$\frac{dF}{da} = \frac{1}{2}\frac{\sigma^2}{\epsilon_0} = \frac{1}{2}\epsilon_0 E^2 , \tag{11.12}$$

and

$$F_1 = -\frac{d^2W}{dz^2}, \tag{11.13}$$

where σ is the surface charge density.

11.3. Examples

We now present several examples that apply to experimental situations encountered frequently in electric force microscopy.

11.3.1. Plane Parallel Capacitor

Figure 11.1 shows two conducting plates with area S separated by a distance z_0 with a dielectric film with thickness z_1 and dielectric constant $\epsilon_0\epsilon_r$ embedded between these two. For $z_1 = 0$, we find that the capacitance, energy, and force are

$$C = \epsilon_0 \frac{S}{z}, \tag{11.14}$$

$$W = \frac{1}{2} CV^2 , \tag{11.15}$$

and

$$F = - \frac{1}{2} C' V^2 , \tag{11.16}$$

respectively, where $C' = \partial C/\partial z$. For $z_1 > 0$, we consider the total capacitance as that of two capacitors in series, obtaining

$$C = \epsilon_0 \frac{S}{z_0 - z_1 + z_1/\epsilon_r} , \tag{11.17}$$

$$W = \frac{1}{2} \epsilon_0 \frac{S}{z_0 - z_1 + z_1/\epsilon_r} V^2 , \tag{11.18}$$

and

$$F = \frac{1}{2} \epsilon_0 \frac{S}{[z_0 - z_1 + z_1/\epsilon_r]^2} V^2 \tag{11.19}$$

for the capacitance, energy, and force, respectively.

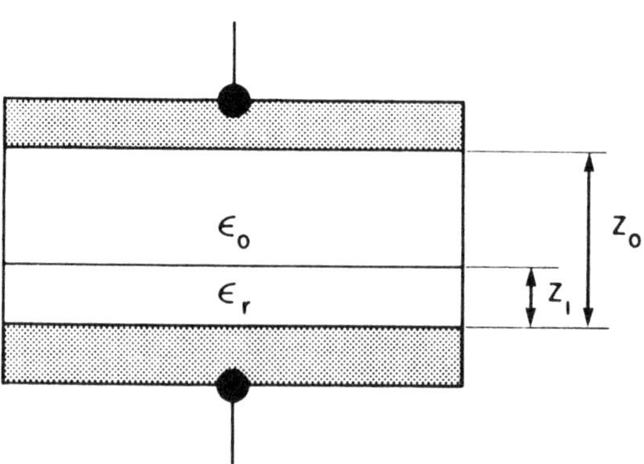

Fig. 11.1 Two conducting plates.

11.3.2. Sphere and Plane

Figure 11.2 shows a conducting sphere with radius R at a distance z from a conducting plate. The capacitance of the isolated sphere, C_0, is

$$C_0 = 4\pi\epsilon_0 R \ , \tag{11.20}$$

and the capacitance of the sphere-plane structure, $C_1 = C - C_0$, where C is the total capacitance, is

$$C_1 = 4\pi\epsilon_0 R \sum_{n=2}^{\infty} \frac{\sinh(\alpha)}{\sinh(n\alpha)} \ , \tag{11.21}$$

with

$$\alpha = \log_e \left[1 + \frac{z}{R} + \sqrt{\frac{z^2}{R^2} + 2\,\frac{z}{R}} \right] . \tag{11.22}$$

For $z/R > 1$, the capacitance converges to

$$C = 2\pi\epsilon_0 R^2/z \ , \tag{11.23}$$

where $2\pi R^2$ is the effective area of the capacitor plate. Figure 11.3 shows C_1 as a function of distance for $R = 1000$ Å in the range $0 < z <$ 5000 Å. The full line refers to the exact value of the capacitance evaluated using Eq. (11.21), while the dashed curve is that of the approximate value of the capacitance using Eq. (11.23). We note that although the approximate solution is valid only for $z/R > 1$, it may sometimes be useful to employ it as a simple analytic tool for describing a given situation.

11.3.3. Uniform Surface Charge Distributed in a Strip

For a uniform surface charge distributed in an infinite strip placed in the x-y plane, shown in Fig. 11.4, the electric field is

$$\mathbf{E} = \frac{\sigma}{2\pi\epsilon_0} [a\mathbf{i}_1 + (b - c)\mathbf{i}_3] \ , \tag{11.24}$$

where $\mathbf{i}_1$ and $\mathbf{i}_3$ are unit vectors in the x and z directions, respectively.

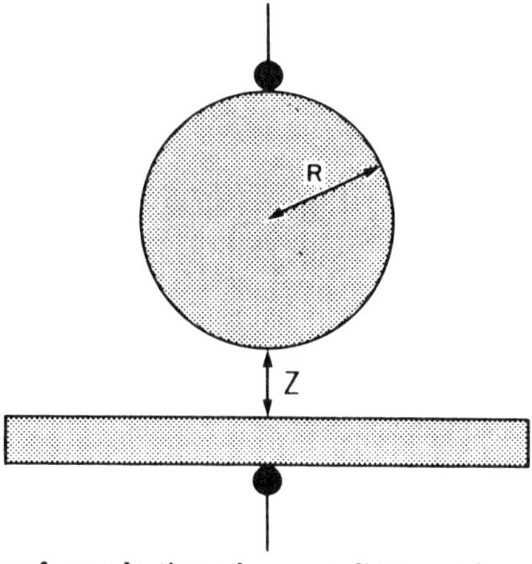

Fig. 11.2 Geometry of a conducting sphere at a distance z from a conducting plate.

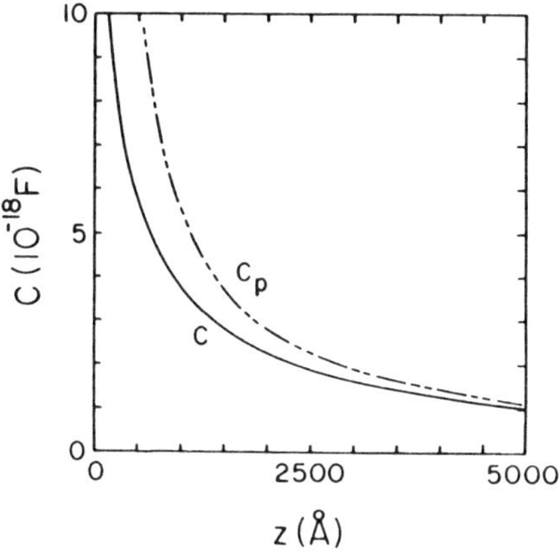

Fig. 11.3 Metal sphere plane exact (C) and approximate (C_p) capacitance as a function of z.

Here

$$a = \ln \sqrt{(x_+{}^2+z^2)/(x_-{}^2+z^2)} \, , \tag{11.25}$$

$$b = atn(x_+/z) \, , \tag{11.26}$$

$$c = atn(x_-/z) \, , \tag{11.27}$$

and

$$x_- = x - d/2 \, , \quad x_+ = x + d/2 \, . \tag{11.28}$$

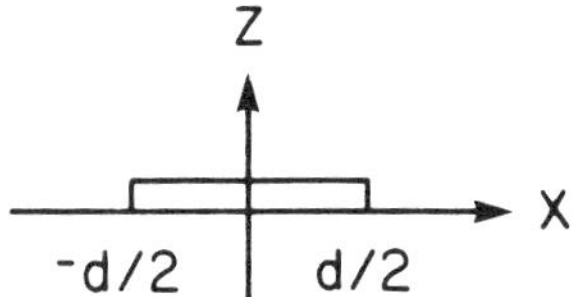

Fig. 11.4 Geometry of a metal strip.

11.3.4. Charged Circular Disk

The field on the axis of symmetry of a circular disk with radius R, shown in Fig. 11.5, whose total charge is q, can be obtained using its surface charge density

$$\sigma = \frac{q}{\pi R^2}, \tag{11.29}$$

yielding

$$E_z(z) = \frac{q}{2\pi\epsilon_0 R^2}[1 - z/\sqrt{R^2 + z^2}] . \tag{11.30}$$

11.3.5. Strips with Alternating Potential

Figure 11.6 shows a structure consisting of metal strips separated from a grounded plate by a distance z_0. The metal strips have alternating potentials $\pm V$ with periodicity p. The solution to the potential is obtained as a Fourier series given by

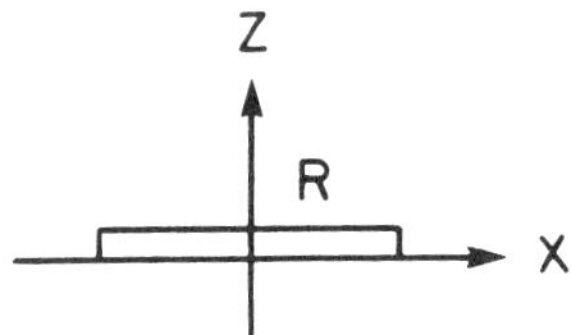

Fig. 11.5 Geometry of a circular metal disk.

$$V(y,z) = \sum_{n=-\infty}^{n=\infty} -\frac{V_0}{\pi i n} [\exp(-i\pi n) - 1] \exp\left(\frac{2\pi i n y}{p}\right) f(\zeta_0,\zeta,n) \,. \qquad (11.31)$$

Here

$$f(\zeta_0,\zeta,n) = \frac{\exp(-n\zeta_0)\exp(n\zeta) - \exp(n\zeta_0)\exp(-n\zeta)}{\exp(-n\zeta_0) - \exp(n\zeta_0)} \,, \qquad (11.32)$$

where $\zeta_0 = 2\pi z_0/p$ and $\zeta = 2\pi z/p$. If the grounded plate is moved far from the strips so that $z_0 >> p$, then the solution becomes

$$V(y,z) = \sum_{n=-\infty}^{n=\infty} -\frac{V_0}{\pi i n} [\exp(-\pi i n) - 1] \exp(2\pi i n y/p) \exp(-|2\pi n z/p|) \,. \qquad (11.33)$$

The first-order solution, using only the term $n = \pm 1$, gives

$$V(y,z) = \frac{4V_0}{\pi} \sin\left(\frac{2\pi y}{p}\right) \exp\left(\frac{-2\pi z}{p}\right) , \qquad (11.34)$$

and the fields are readily obtained by taking the gradient of these potentials.

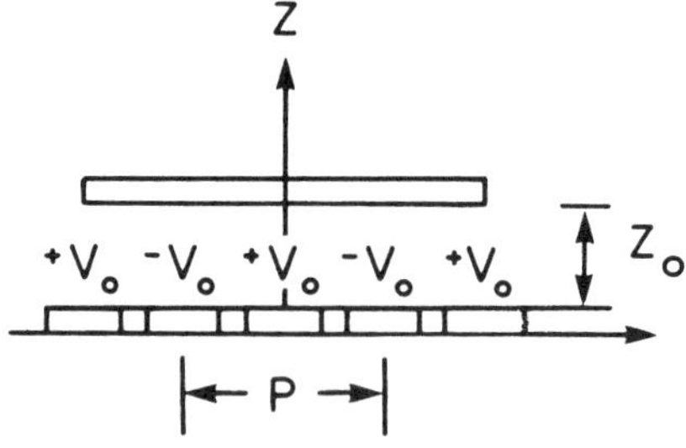

Fig. 11.6 Geometry of metal strips at a distance z from a conducting plate.

11.4. Principles of Operation

In this section, we develop the theory of operation of electric force microscopy for systems using a bimorph-driven lever, a sample-driven lever, and a voltage-driven lever. We derive general expressions that give the change in amplitude of vibration of the lever supporting the force-sensing conducting tip in terms of the parameters of the system and discuss the signal-to-noise ratio. The amplitude of vibration will be a function of the electric potentials, the topography, the dielectric constant, and the charges across the surface of the sample. We can then raster scan the tip across the sample using any of the systems described in Chapters 4 through 10 and obtain a mapping of these properties. The forces with which we are concerned will usually have derivatives in the direction of vibration of the lever, and can therefore be written as $F = F_0 + F_1\varsigma$ where F_1 and ς are the force derivative and the amplitude of vibration of the lever, respectively.

11.4.1. Bimorph-Driven Lever

Figure 11.7 shows a schematic of a typical system where the lever is mounted on a bimorph vibrating with amplitude a and frequency ω. In this configuration we have for u, z, and g,

$$u = u_0 + a \exp(i\omega t) \, , \tag{11.35}$$

$$z = z_0 + \varsigma \, , \tag{11.36}$$

$$\varsigma = A_b(\omega) \exp[i(\omega t - \theta)] \, , \tag{11.37}$$

and

$$g = 0 \, . \tag{11.38}$$

Here A_b and ω are the amplitude and frequency of vibration of the lever and θ is a phase angle. The equation of motion for this system is

$$m \frac{\partial^2 z}{\partial t^2} + \frac{\omega_0 m}{Q} \frac{\partial z}{\partial t} + k(z - u) = F(z) \, , \tag{11.39}$$

where Q is the quality factor of the lever, m its mass, and ω_0 its free-motion resonance frequency. To find $F(z)$ for this case, we consider external voltages V_{dc} and $V_{ac} = V_1\sin(\Omega t)$ applied between the lever and sample together with a charge q_s deposited on a thin insulating film placed on top of the sample. As explained in the section dealing with

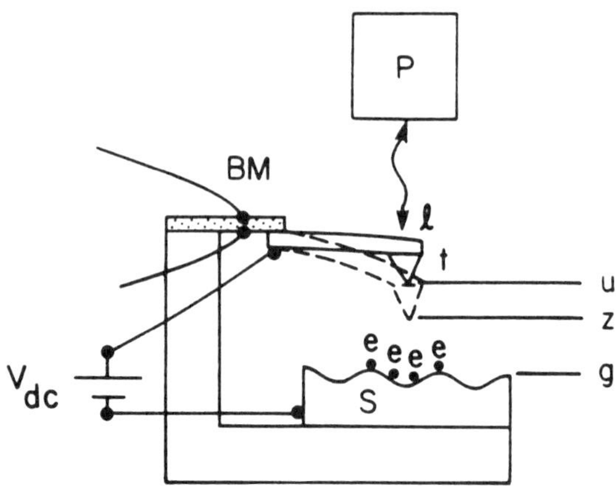

Fig. 11.7 Bimorph-driven lever with charged sample and applied potentials.

applications, Ω will be chosen to be much smaller than ω_0. Now, the simplest approach is to assume that the voltages V_{dc} and V_{ac} induce charges on the tip given by

$$q_{dc} = CV_{dc} \tag{11.40}$$

and

$$q_{ac} = CV_{ac} , \tag{11.41}$$

respectively, with a total charge given by

$$q_t = q_s + q_{dc} + q_{ac} . \tag{11.42}$$

The force acting on the tip will then be due to the charge-charge interaction and the capacitive energy,

$$F = \frac{q_s q_t}{4\pi\epsilon_0 z^2} + \frac{1}{2}C'(V_{dc} + V_{ac})^2 . \tag{11.43}$$

Substituting for the total charge gives

$$F = \frac{1}{4\pi\epsilon_0 z^2} [q^2_s + q_s q_{dc} + q_s q_{ac}] + \frac{1}{2}C'[V_{dc} + V_{ac}]^2 . \tag{11.44}$$

We now decompose the force into components at dc and at the frequencies Ω and 2Ω,

$$F_{dc} = \frac{q^2_s}{4\pi\epsilon_0 z^2} + \frac{q_s V_{dc}}{4\pi\epsilon_0}\frac{C}{z^2} + \frac{1}{2}C'\left[V^2_{dc} + \frac{1}{2}V^2_1\right], \tag{11.45}$$

$$F_\Omega = \left[\frac{q_s V_1}{4\pi\epsilon_0}\frac{C}{z^2} + C'V_{dc}V_1\right]\sin(\Omega t), \tag{11.46}$$

and

$$F_{2\Omega} = -\frac{1}{4}C'V^2_1\cos(2\Omega t). \tag{11.47}$$

The force derivatives, which are readily obtained from the force, are

$$F_{1dc} = -\frac{2q^2_s}{4\pi\epsilon_0 z^3} + \frac{q_s V_{dc}}{4\pi\epsilon_0}\frac{C'z^2-2Cz}{z^4} + \frac{1}{2}C''\left[V^2_{dc} + \frac{1}{2}V_1^2\right], \tag{11.48}$$

$$F_{1\Omega} = \left[\frac{q_s V_1}{4\pi\epsilon_0}\frac{C'z^2-2Cz}{z^4} + C''V_{dc}V_1\right]\sin(\Omega t), \tag{11.49}$$

and

$$F_{12\Omega} = -\frac{1}{4}C''V^2_1\cos(2\Omega t). \tag{11.50}$$

The average position of the lever will be determined by equating the electrostatic and restoring forces, yielding

$$k(z_0 - u_0) = F(z_0). \tag{11.51}$$

The effective spring constant, which is given in terms of the derivative of the force by

$$k' = k - \langle F_1(z_0)\rangle, \tag{11.52}$$

shifts the resonance frequency of the lever from

$$\omega_0 = \sqrt{k/m} \tag{11.53}$$

to

$$\omega'_0 = \sqrt{k'/m}. \tag{11.54}$$

We note that the feedback-driven lever, which operates at $\omega = \omega'_0$, directly measures the force derivative from the frequency shift,

$$F_1 = 2k \frac{\delta\omega}{\omega_0} . \qquad (11.55)$$

The amplitude of vibration of the bimorph-driven lever is given by

$$A_b(\omega) = a \frac{k}{k'} \frac{Q}{[Q^2(1 - \omega^2/\omega'^2_0)^2 + \omega^2\omega_0^2/\omega'^4_0]^{1/2}} , \qquad (11.56)$$

which, at $\omega = \omega'_0$, is

$$A_b(\omega_0') = aQ \sqrt{k/k'} . \qquad (11.57)$$

The variation of the amplitude of vibration caused by a variation in the force derivative, while operating at the steepest slope of the resonance curve, is now given by

$$\delta A_b(\omega{=}\omega_m) = \frac{2}{3\sqrt{3}} \frac{Q}{k} A_b(\omega = \omega'_0)\, \delta(F_{1dc} + F_{1\Omega} + F_{12\Omega}) . \qquad (11.58)$$

Note that the tip-sample capacitance is determined by the geometries of the tip and the sample as well as their distance, and that a good approximation is the use of a spherical tip, Eq. (11.21).

11.4.2. Sample-Driven Lever

Figure 11.8 shows a schematic of a typical system where the sample is mounted on a bimorph vibrating with amplitude a and frequency ω. As in the previous case, we apply an external voltage, $V_{dc} + V_{ac}$, between the lever and sample and a charge, q_s, deposited on a thin insulating film placed on top of the sample. In this configuration we have for u, z, ζ, and g,

$$u = u_0 , \qquad (11.59)$$

$$z = z_0 + \zeta , \qquad (11.60)$$

$$\zeta = A_s(\omega) \exp[i(\omega t - \theta)] , \qquad (11.61)$$

and

$$g = g_0 + a \exp(i\omega t) , \qquad (11.62)$$

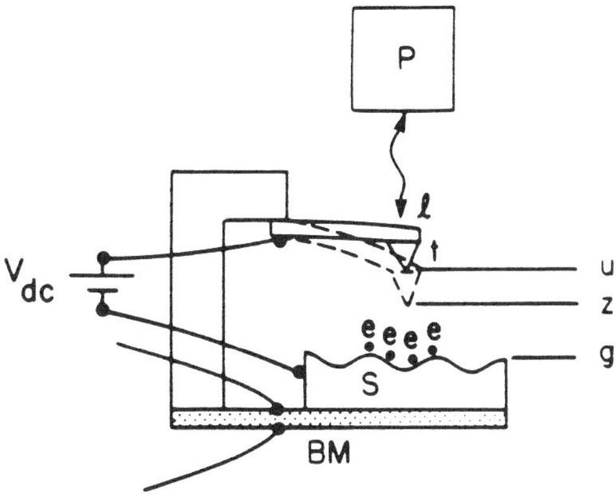

Fig. 11.8 Sample-driven lever with charged sample and applied potentials.

where A_s and ω are the amplitude and frequency of vibration of the lever and a is the amplitude of vibration of the sample. The equation of motion in this case is

$$m \frac{\partial^2 z}{\partial t^2} + \frac{\omega_0 m}{Q} \frac{\partial z}{\partial t} + k(z - u) = F(z - g) . \tag{11.63}$$

The average position of the lever will be determined by equating the electrostatic and restoring forces, yielding

$$k(z_0 - u_0) = F(z_0) , \tag{11.64}$$

where we have set g_0 to zero. The effective spring constant of the lever shifts its resonance frequency to

$$\omega'_0 = \sqrt{k'/m} , \tag{11.65}$$

with an amplitude of vibration at resonance given by

$$A_s(\omega = \omega'_0) = a_s Q \frac{|F_1|}{\sqrt{kk'}} . \tag{11.66}$$

We now linearize Eq. (11.66), getting

$$A_s(\omega = \omega'_0) = a_s Q \frac{|F_1|}{k} \left[1 + \frac{F_1}{2k}\right] . \tag{11.67}$$

The derivatives of the amplitude with respect to the force derivative, on resonance and at the steepest slope of the resonance curve, are given by

$$\frac{\partial A_s(\omega = \omega'_0)}{\partial F_1} = a_s \frac{Q}{k} + a_s Q \frac{F_1}{k^2} \tag{11.68}$$

and

$$\frac{\partial A_s(\omega = \omega_m)}{\partial F_1} = \frac{2}{3\sqrt{3}} a_s \frac{Q^2}{k^2} F_1 , \tag{11.69}$$

respectively. By taking the ratio of Eqs. (11.68) and (11.69), we finally get

$$\frac{\delta A_s(\omega = \omega'_0)}{\delta A_s(\omega = \omega_m)} = \frac{3\sqrt{3}}{2} \frac{k + F_1}{QF_1} \simeq \frac{3\sqrt{3}}{2} \frac{k}{QF_1} . \tag{11.70}$$

If, as is usually the case, $k > QF_1$, then it is advantageous to operate on resonance where $\omega = \omega'_0$.

11.4.3. Voltage-Driven Lever

Figure 11.9 shows a schematic of a system where the lever is driven by applying a voltage V_{dc} and a voltage V_{ac} with frequency Ω between the tip and sample. We also consider the possibility of having an additional dc voltage that results from a charge q_s placed between the tip and sample. In this configuration we have for u, z, ζ, and g,

$$u = u_0 , \tag{11.71}$$

$$z = z_0 + \zeta , \tag{11.72}$$

$$\zeta = A_v(\omega) \exp[i(\omega t - \theta)] , \tag{11.73}$$

and

$$g = 0 , \tag{11.74}$$

where A_v and ω are the amplitude and frequency of vibration of the lever. The equation of motion for this system is

$$m \frac{\partial^2 z}{\partial t^2} + \frac{\omega_0 m}{Q} \frac{\partial z}{\partial t} + k(z-u) = F + \langle F_1 \rangle \zeta . \tag{11.75}$$

The average position of the lever will be determined by equating the electrostatic and restoring forces, yielding

$$k(z_0 - u_0) = F(z_0) . \tag{11.76}$$

We will denote the vibration of the lever at the frequencies $\omega = \Omega$ and $\omega = 2\Omega$ by

$$\zeta_\Omega = A_v (\omega = \Omega) \sin(\Omega t + \phi_1) , \tag{11.77}$$

and

$$\zeta_{2\Omega} = A_v (\omega = 2\Omega) \sin(2\Omega t + \phi_2) . \tag{11.78}$$

For $\omega = \Omega$, we get the equation of motion

$$\left[k' - \omega^2 m + i \frac{m\omega\omega_0}{Q} \right] \zeta_\Omega = F_\Omega , \tag{11.79}$$

resulting in an amplitude of vibration given by

$$A_v (\omega = \Omega) = \frac{|F_\Omega|}{k'} \frac{Q}{[Q^2[1 - \omega^2/\omega_0'^2]^2 + \omega^2\omega_0^2/\omega_0'^4]^{1/2}} . \tag{11.80}$$

On resonance, we get

$$A_v (\omega = \Omega = \omega'_0) = Q \frac{|F_\Omega|}{\sqrt{k(k - F_{1dc})}} \tag{11.81}$$

for the amplitude of vibration, which, on linearization yields

$$A_v (\omega = \Omega = \omega'_0) = \frac{Q}{k} |F_\Omega| + \frac{1}{2} \frac{Q}{k^2} |F_\Omega| F_{1dc} . \tag{11.82}$$

The derivative of the amplitude with respect to the force, for $\omega = \Omega = \omega'_0$, is

$$\frac{\partial A_v(\omega = \Omega = \omega'_0)}{\partial F_\Omega} = \frac{Q}{k} + \frac{1}{2}\frac{Q}{k^2} F_{1dc} \simeq \frac{Q}{k}. \tag{11.83}$$

The derivatives of the amplitude with respect to the force derivative, for $\omega = \Omega = \omega'_0$ and $\omega = \Omega = \omega_m$, are

$$\frac{\partial A_v(\omega = \Omega = \omega'_0)}{\partial F_{1dc}} = \frac{1}{2}\frac{Q}{k^2} F_\Omega \tag{11.84}$$

and

$$\frac{\partial A_v(\omega = \Omega = \omega_m)}{\partial F_{1dc}} = \frac{2}{3\sqrt{3}}\frac{Q^2}{k^2} F_\Omega , \tag{11.85}$$

respectively. Taking the ratio of Eqs. (11.84) and (11.85) finally yields

$$\frac{\delta A_v(\omega = \Omega = \omega'_0)}{\delta A_v(\omega = \Omega = \omega_m)} = \frac{3\sqrt{3}}{4Q}. \tag{11.86}$$

Since $Q >> 1$, we find that it is advantageous to operate at $\omega = \omega_m$ for measuring force derivatives, while for the measurement of forces it is best to operate at $\omega = \omega'_0$. We note that Eqs. (11.79) through (11.86) apply to the case $\omega = 2\Omega$ as well.

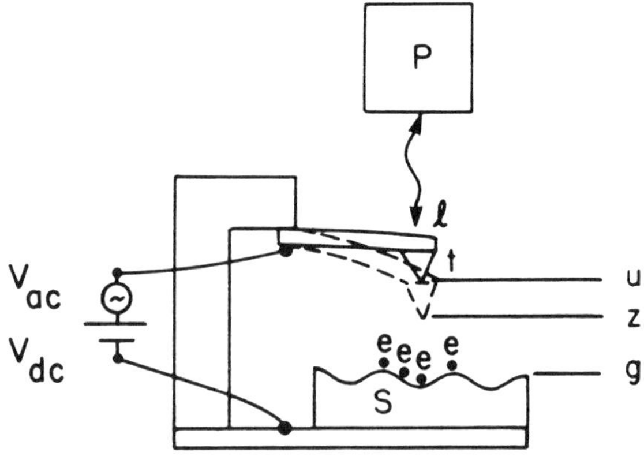

Fig. 11.9 Voltage-driven lever with charged sample and applied potentials.

11.5. Noise Considerations

We found that the thermal noise of the lever is given by

$$\langle \delta z^2 \rangle_\ell = \frac{4KTBQ}{k\omega_0} . \qquad (11.87)$$

We will now apply this result to our three cases and find the minimum value of the force derivative δF_1 that is limited by lever noise.

11.5.1. Bimorph-Driven Lever

For the bimorph-driven lever, operating at $\omega = \omega_m$, we get

$$\delta F_{1min} = \frac{3\sqrt{3}}{2} \frac{k}{aQ^2} \langle \delta z^2 \rangle^{1/2} . \qquad (11.88)$$

For the feedback-driven lever, operating at $\omega = \omega'_0$, the natural linewidth of the noise-generated Lorentzian is

$$\Delta\omega = \frac{\omega_0}{Q} . \qquad (11.89)$$

In the presence of amplification, the line narrows to $\Delta\omega'$,

$$\Delta\omega' = \frac{\omega_0}{Q} \frac{KT}{kA^2} , \qquad (11.90)$$

with an effective quality factor Q',

$$Q' = \frac{kA^2}{KT} Q . \qquad (11.91)$$

We can define a minimum detectable change in the oscillating frequency as that which is equal, say, to the natural linewidth, yielding

$$\delta F_{1min} = 2k \frac{\Delta\omega'}{\omega_0} = \frac{2KT}{QA^2} = \frac{2k}{Q'} . \qquad (11.92)$$

11.5.2. Sample-Driven Lever

Here we find, for operation at $\omega = \omega'_0$,

$$\delta F_{1min} = \frac{k}{aQ} \langle \delta z^2 \rangle^{1/2} . \tag{11.93}$$

11.5.3. Voltage-Driven Lever

Here we have, for operation at $\omega = \omega'_0$ and at $\omega = \omega_m$, respectively,

$$\delta F_{1min} = \frac{2k^2}{QF} \langle \delta z^2 \rangle^{1/2} \tag{11.94}$$

and

$$\delta F_{1min} = \frac{3\sqrt{3}}{2} \frac{k^2}{Q^2 F} \langle \delta z^2 \rangle^{1/2} . \tag{11.95}$$

11.6. Applications

In this section we use the bimorph-driven lever case as an example and derive expressions for δF, δz, $\delta \epsilon$, δq, and δV in terms of δA,

$$\delta A_b(\omega = \omega_m) = \frac{2}{3\sqrt{3}} \frac{aQ^2}{k} \delta F_1 . \tag{11.96}$$

Expressions for the sample-driven lever and voltage-driven lever methods are straightforward.

11.6.1. Modeling the Force

The force can be obtained using

$$\delta F_{1dc} = \frac{3\sqrt{3}}{2} \frac{k}{aQ^2} \delta A(\omega = \omega_m) . \tag{11.97}$$

11.6.2. Modeling the Topography

Here we assume that $q_s = 0$ and $V_1 = 0$, and for simplicity, that $C = 2\pi\epsilon_0 R^2/z$. The second derivative of the capacitance

$$C'' = \frac{4\pi\epsilon_0 R^2}{z^3} \tag{11.98}$$

gives for the variation in the force derivative,

$$\delta F_1 = \frac{6\pi\epsilon_0 R^2 V^2{}_{dc}}{z^4}\,\delta z\ . \tag{11.99}$$

Equating Eq. (11.99) with Eq. (11.97) gives

$$\delta z = \frac{\sqrt{3}}{4\pi\epsilon_0}\,\frac{kz^4}{aQ^2R^2V^2{}_{dc}}\,\delta A\ . \tag{11.100}$$

11.6.3. Modeling the Dielectric Constant

From the second derivative of the capacitance and the force derivative, we get the variation in the force derivative

$$\delta F_1 = \frac{2\pi R^2 V^2{}_{dc}}{z^3}\,\delta\epsilon\ , \tag{11.101}$$

yielding for $\delta\epsilon$

$$\delta\epsilon = \frac{3\sqrt{3}}{4\pi}\,\frac{kz^3}{aQ^2R^2V^2{}_{dc}}\,\delta A\ . \tag{11.102}$$

11.6.4. Modeling the Charge

Here we set V_{dc} to zero and get the constant force derivative

$$F_{1dc} = -\frac{q^2{}_s}{2\pi\epsilon_0 z^3} + \frac{1}{4}C''V_1{}^2\ , \tag{11.103}$$

and the component at Ω,

$$F_{1\Omega} = \left[\frac{q_s V_1}{4\pi\epsilon_0}\,\frac{C'z^2-2Cz}{z^4}\right]. \tag{11.104}$$

The assumption is that Ω is much smaller than ω_m, yet larger than the response of the feedback electronics controlling the tip-sample distance. Consequently, only the constant component of the force derivative will affect the electronic feedback. The component at frequency Ω, however, will give a signal, detectable with a phase-sensitive detector, that is proportional to the magnitude and sign of the charge q_s (Terris et al. 1989a,b).

11.6.5. Modeling the Voltage

The voltage across a sample can be obtained using the second derivative of the capacitance and the force derivative,

$$\delta F_1 = \frac{4\pi\epsilon_0 R^2 V_{dc}}{z^3}\,\delta V_{dc}\ , \tag{11.105}$$

yielding

$$\delta V_{dc} = \frac{3\sqrt{3}}{8\pi\epsilon_0}\,\frac{kz^3}{aQ^2R^2V_{dc}}\,\delta A\ . \tag{11.106}$$

11.7. Performance

We present a summary of the few experimental results that have been interpreted using models that are analogous to those presented in this chapter.

11.7.1. Measurement of Forces

Martin et al. (1988b) measured the electrostatic force as a function of tip-to-sample distance using bimorph-driven-lever and voltage-driven-lever methods. Modeling the capacitance as that of a plane-parallel capacitor, they obtained good agreement between theory and experiment for distances ranging from several nm to 170 nm, and forces ranging up to 12 nN, where the amplitude of vibration ranged from 0.1 Å to 5 Å. They find that electrostatic forces as small as 10^{-10} N can be measured. Terris et al. (1989a,b) measured the force gradient as a function of distance up to a distance of 1 μm and obtained an agreement with the theory that models the tip as a conducting sphere. Schönenberger and Alvarado (1989, 1990a) measured forces in the range of 14×10^{-9} N as a function of tip-to-sample distance in the range of 200-nm, for potentials on the order of several volts.

11.7.2. Measurement of Topography

Anders and Heiden (1988, 1990) imaged Nb_3Sn films to a distance of 2000 Å. Martin et al. (1988b) measured the topography across *a* Si wafer that was partially covered with a 1-μm-thick photoresist using the voltage-driven-lever method. Schönenberger et al. (1990) measured the topography of grooves in an etched glass coated with a gold plating.

11.7.3. Measurement of Dielectric Constant

For a capacitor that is partially filled with a dielectric film, we can measure the change in capacitance, which is a function of both the film thickness and its dielectric constant. Martin et al. (1988b) measured capacitance across *a* Si wafer, which was partially covered with a 1-μm-thick photoresist using the voltage-driven-lever method. They find that a capacitance as small as 10^{-19} F can be measured.

11.7.4. Measurement of Charge

Stern et al. (1988), Terris et al. (1989a,b, 1990a,b), Barrett and Quate (1991, 1992), Schönenberger and Alvarado (1990b) measured deposited localized charges on an insulator placed between the tip and sample as shown in Figs. 11.10 and 11.11. The images were proven to result from the charges by varying the tip-sample dc voltage and observing the effect on the image. They found that their system was sensitive to the charge of six electrons. Ferroelectric domains have been imaged by Saurenbach and Terris (1990).

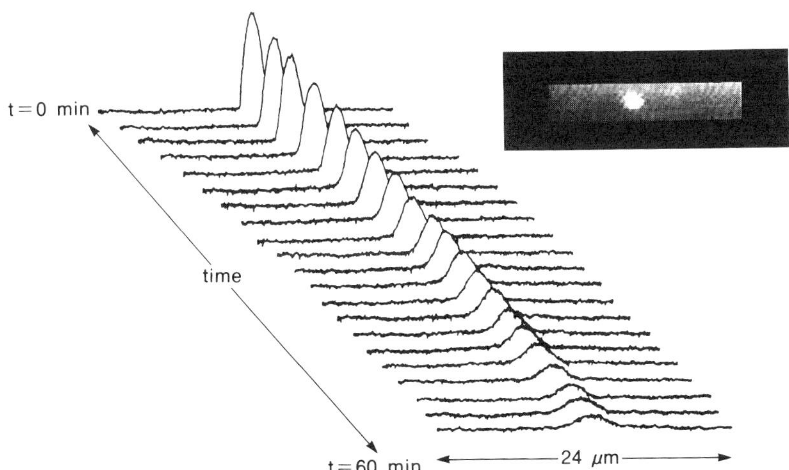

Fig. 11.10 Electric force microscope image of a charge deposited on PMMA in the process of decay. (Courtesy: J. E. Stern, B. D. Terris, H. J. Mamin, and D. Rugar, IBM.)

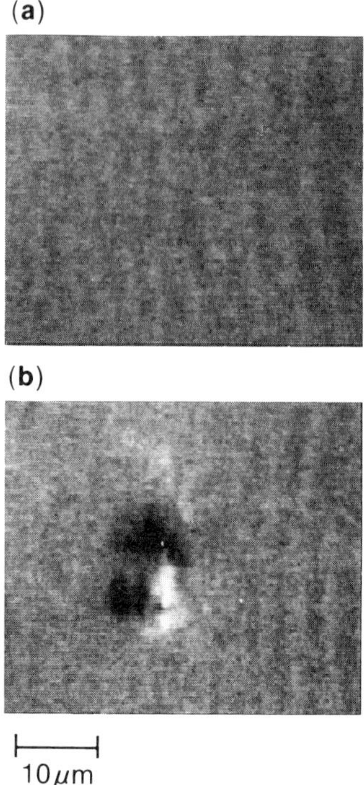

Fig. 11.11 Electric force microscope image of positive and negative charge deposited on PMMA. (Courtesy: B. D. Terris, J. E. Stern, D. Rugar, and H. J. Mamin, IBM.)

11.7.5. Measurement of Potential

Muralt and Pohl (1986) and Muralt et al. (1987) measured the potential across a Schottky barrier and a GaAs heterostructure using scanning tunneling microscopy. Subsequently, Martin et al. (1988b), Weaver and Abraham (1990), and Weaver and Wickramasinghe (1990), using electric force microscopy, measured the potential across a pn junction of a commercial transistor, Morita et al. (1989) measured surface conductance of metals, contact potentials (Kelvin probes) were measured by Nonnenmacher et al. (1992), and impurity dopant density was measured by Huang et al. (1992).

11.8. Summary

Table 11.1 gives the system, interaction, frequency of operation, and working expression for the different systems discussed in this chapter. The exact expression for the forces and their derivatives, in the presence of charges and externally applied potentials, are given by Eqs. (11.44) through (11.50).

Table 11.1 Characterization of the bimorph-driven (BM), feedback-driven (FB), sample-driven (S), and voltage-driven (V) systems.

System	Interaction	Frequency	Expression
BM	F_1	ω_m	$\delta A = \dfrac{2}{3\sqrt{3}} \dfrac{aQ^2}{k} \delta F_1$
FB	F_1	ω'_0	$\delta\omega = \dfrac{\omega_0}{2k} \delta F_1$
S	F_1	ω'_0	$\delta A = \dfrac{aQ}{k} \delta F_1$
V	F_1	ω_m	$\delta A = \dfrac{2Q^2}{3\sqrt{3}k^2} F \, \delta F_1$
V	F	ω'_0	$\delta A = \dfrac{Q}{k} \delta F$

12
Magnetic Force Microscopy

12.1. Introduction

Chapter 12 describes the principles of operation of magnetic force microscopy, which maps magnetostatic forces across the surface of a sample. Note that the interaction between a tip and a magnetic sample can also be probed using the STM (Allenspach et al. 1987, Allenspach and Bischof 1989, Wiesendanger et al. 1990, 1992, Moreland and Rice 1990b, 1991, Yi et al. 1992, DiCarlo et al. 1992, and Watanuki et al. 1992). These forces, which can be either attractive or repulsive, act between the magnetic tip supported by the flexible lever and the magnetic sample. We start with the basic theory of magnetostatics and give several examples that represent relevant cases. These include the interaction of two dipoles, the field of a uniformly magnetized sphere, and the interaction of the tip with magnetic lines written on storage media. Next we develop the theory of operation of scanning force microscopy and derive general expressions for bimorph-driven, sample-driven, and voltage-driven lever systems. The amplitude of vibration of the lever is obtained in terms of the parameters of the system and the resultant signal-to-noise ratio is discussed. We then present experimental results where polar and longitudinal magnetic domains across a sample, and magnetic fields across a recording head have been mapped. We conclude with a summary that gives working equations for each system. For recent developments, see for example Grütter et al. (1992), Sidles et al. (1992), Rugar et al. (1992), Volodin and Marchevsky (1992), and Göddenhenrich et al. (1992).

12.2. Basic Concepts

In this section we present basic concepts of magnetostatics and, in particular, the interaction between two magnetized media. In contrast to electrostatics, where a few electrons can already be detected by electric force microscopy, the interaction here derives from cooperative phenomena that requires the participation of a large number of atoms of a ferromagnetic medium. As a result, the theory of magnetic force microscopy is more involved than that of electric force microscopy. In particular, the images obtained by magnetic force microscopy result from a convolution of the magnetic properties of both the sample and the probing tip, which in most cases are not known exactly. While the role of the basic field in electrostatics is the **E** field, here it is the magnetic induction **B**, measured in units of tesla or weber/meter². Since there are no magnetic monopoles, we get from Maxwell's equations that

$$\nabla \cdot \mathbf{B} = 0 \,, \tag{12.1}$$

where **B** is the magnetic flux density. We can define a vector potential **A** from which **B** is derived using

$$\mathbf{B} = \nabla \times \mathbf{A} \,. \tag{12.2}$$

Since the magnetic media of interest are macroscopic, it will be convenient to present their total magnetic dipole moment **m** in terms of the magnetization **M** by

$$\mathbf{m} = \int_V \mathbf{M} dv \,. \tag{12.3}$$

The magnetization can be described in terms of bulk and surface equivalent current densities, $\mathbf{J}_e$ and $\boldsymbol{\alpha}_e$, given, respectively, by

$$\mathbf{J}_e = \nabla \times \mathbf{M} \tag{12.4}$$

and

$$\boldsymbol{\alpha}_e = \mathbf{M} \times \mathbf{n} \,, \tag{12.5}$$

where **n** is a unit vector normal to the surface. Using Eqs. (12.4) and (12.5), we can write the potential **A** as

$$\mathbf{A} = \frac{\mu_0}{4\pi} \int_S \frac{\boldsymbol{\alpha}_e}{r} dS + \frac{\mu_0}{4\pi} \int_v \frac{\mathbf{J}_e}{r} dv \,. \tag{12.6}$$

Here the permeability μ, measured in newton/ampere² or in henry/meter, has a value of $4\pi \times 10^{-7}$ for free space. The sources of magnetic induction **B** are the free and equivalent current densities, $\mathbf{J}_e$ and $\mathbf{J}_f$, respectively. These current densities determine the curl of **B** through

$$\nabla \times \mathbf{B} = \mu_0 \left[\mathbf{J}_f + \mathbf{J}_e \right] . \tag{12.7}$$

Substituting for the equivalent current density using Eq. (12.4) gives

$$\nabla \times \mathbf{B} = \mu_0 [\mathbf{J}_f + \nabla \times \mathbf{M}] \tag{12.8}$$

and

$$\nabla \times \left[\frac{\mathbf{B}}{\mu_0} - \mathbf{M} \right] = \mathbf{J}_f . \tag{12.9}$$

Equation (12.9) leads to the definition of the magnetic field **H**,

$$\mathbf{H} = \frac{\mathbf{B}}{\mu_0} - \mathbf{M} , \tag{12.10}$$

measured in ampere/meter. In regions where $\mathbf{J}_f = 0$, it is possible to define a scalar potential Φ from which **H** can be derived using

$$\mathbf{H} = - \nabla\Phi . \tag{12.11}$$

Instead of using equivalent bulk and surface current densities, we can define magnetic bulk and surface "charges," ρ_m and σ_m, given, respectively, by

$$\rho_m = -\nabla \cdot \mathbf{M} \tag{12.12}$$

and

$$\sigma_m = \mathbf{M} \cdot \mathbf{n} . \tag{12.13}$$

The potential Φ, in terms of these "charges," is

$$\Phi = \frac{1}{4\pi} \int_V \frac{\rho_m}{r} dV + \frac{1}{4\pi} \int_S \frac{\sigma_m}{r} dS \tag{12.14}$$

or

$$\Phi = -\frac{1}{4\pi} \int_V \frac{\nabla \cdot \mathbf{M}}{r} dV + \frac{1}{4\pi} \int_S \frac{\mathbf{M} \cdot \mathbf{n}}{r} dS . \tag{12.15}$$

Here V and S are the volume and surface of integration. We will be concerned with forces that an external magnetic induction **B** exerts on a medium having a magnetic dipole moment **m**. This force is given by

$$\mathbf{F} = (\mathbf{m} \times \nabla) \times \mathbf{B} = (\mathbf{m} \cdot \nabla)\, \mathbf{B} = \nabla\, (\mathbf{m} \cdot \mathbf{B}) , \tag{12.16}$$

where we have used vector identities to derive the right-hand side expression. The torque τ exerted by this force and the interaction energy are given, respectively, by

$$\tau = \mathbf{m} \times \mathbf{B} \tag{12.17}$$

and

$$W = -\mathbf{m} \cdot \mathbf{B} \ . \tag{12.18}$$

Note that to describe the force and interaction energy of a system consisting of a probing tip and a sample, we have to perform integrations across the volume and surface of each one using Eqs. (12.16) and (12.18).

12.2.1. Magnetic Media

Figure 12.1 shows the hysteresis loop of a magnetic medium where the magnetization M is plotted as a function of the applied external field H. The important points on this loop are the remanent magnetization (remanence) M_r [A/m], the saturation magnetization M_s [A/m], and the coercivity (coercive force) H_c [A/m]. M_s is a measure of the maximum magnetization that the external field can generate in the given material, while M_r denotes the value of the magnetization after the field has been removed. The coercivity gives the value of the external field required to bring the magnetization to zero. The literature sometimes presents values pertaining to the saturation magnetization or remanence in terms of the magnetic intensity $\mathbf{I} = \mu_0 \mathbf{M}$, in units of tesla. Tables 12.1, 12.2, and 12.3 present parameters relating to the two possible recording orientations. In the longitudinal orientation (l), the domains have a magnetization aligned in the plane of the surface of the storage medium. In the perpendicular orientation (p), which allows for a higher recording density, the domains are oriented perpendicular to the surface of the storage medium.

12.3. Examples

In this section we present several examples relevant to experiments where magnetic force microscopy is used to image magnetostatic fields across the surface of a sample.

12.3.1. Dipole-Dipole Interaction

The interaction energy of two point dipoles with magnetic moments $\mathbf{m}_i$ is given by

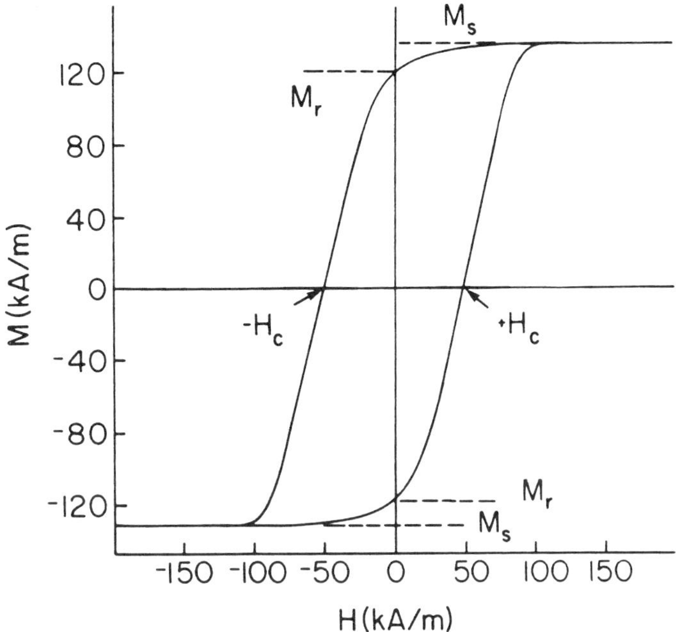

Fig. 12.1 Typical hysteresis loop of a ferromagnet.

$$W = \frac{\mu_0}{4\pi}\left[\frac{\mathbf{m}_1\cdot\mathbf{m}_2}{r^3} + \frac{3(\mathbf{m}_1\cdot\mathbf{r})(\mathbf{m}_2\cdot\mathbf{r})}{r^5}\right], \tag{12.19}$$

where $\mathbf{r}$ is the vector connecting the two dipoles.

12.3.2. Uniformly Magnetized Sphere

The magnetic moment of a uniformly magnetized sphere with radius R is

$$\mathbf{m} = \frac{4\pi}{3} R^3 \mathbf{M} . \tag{12.20}$$

Inside the sphere, the magnetic field is antiparallel to the magnetic induction, which itself is parallel to the magnetization, while outside the sphere, the magnetic field is parallel to the magnetic induction. The magnetic induction of the uniformly magnetized sphere is readily found to be

Table 12.1 Parameters relating to longitudinal (l) and perpendicular (p) magnetic recording materials (Ullmann's Encyclopedia of Industrial Chemistry, 1989).

Material	Curie T [^{0}C]	saturation M_s [kA/m]	coercivity H_c [kA/m]	particle ℓ [μm]	axial ratio
γ-Fe_2O_3 (l)	590	278-358	20-35	0.3-1.0	3-10
γ-Fe_2O_3-Co (l)	525	318-350	40-75	0.17-0.7	3-10
α-Fe (l)	770	716-1034	80-120	0.15-0.5	4-7
CrO_2 (l)	117	334-374	40-62	0.15-0.4	10-15
Co-P (l)	$\simeq$1000	636-954	40-80		
Co-Ni (l)	$\simeq$1000	636-1273	40-100		
$BaFe_{12}O_{19}$ (p) Ti doped	350	110	72	0.08	
$BaFe_{12}O_{19}$ (p) Ti+Co doped	350	114	76	0.085	
Co-Cr (p)		300-700	20-100		

l =longitudinal recording
p =perpendicular recording
ℓ =length

$$\mathbf{B} = \frac{\mu_0}{4\pi}\left[-\frac{\mathbf{m}}{r^3} + \frac{3(\mathbf{m}\cdot\mathbf{r})\mathbf{r}}{r^5}\right], \tag{12.21}$$

where $\mathbf{r}$ is a unit vector pointing from one dipole to the other. Here $\mathbf{B}$ is identical to that of a magnetic dipole, a result which is useful in modeling the interaction between a magnetic tip and magnetic domains of a sample.

12.3.3. Bloch Wall Contrast

Hartmann (1988) and Hartmann and Heiden (1988) calculated the force derivative acting between a magnetic sample and a magnetic tip in the vicinity of a 180^0 Bloch wall and obtained an analytic solution for this interaction. Here the tip is modeled as a pyramid with angle θ and height ℓ. The vertical component of the stray magnetic field of a Bloch wall, $H_z(x,y)$, is

$$H_z(x,z) = \frac{1}{\pi} M_s \frac{\mathrm{atn}\,(w/2z)}{\cosh\,(qx/\delta)}, \tag{12.22}$$

Table 12.2 Parameters relating to longitudinal magnetic recording media (Ullmann's Encyclopedia of Industrial Chemistry, 1989).

Application	Carrier	bit ℓ [μm]	remanence M_r [mT]	coercivity H_c kA/m	material
audio	T	2.2	110	27	1X
audio	CCI	1.2	140	30	1X
audio	CCII	1.2	160	45	2X,4x
audio	CCIV	1.2	300	85	3x
DAT	T	0.3	250	120	3X
Video	T	0.5	140-170	50	2X,4X
Video	T	0.3	250	120	3X
S-VHS	T	0.3	160	70	2X
ED-BETA	T	0.3	250	120	3X
data	T	0.6	170	50	2X,4X
data	T	0.6	110		1X
data	FD	0.6-2.5	75		2X
data	FD	0.3	180	100	3X
data	RD	4	75	27	1X
data	RD	1.4	75	50	2X
data	RD	0.7	500	100	5X

T =tape
CC =compact cassette tape
FD =flexible disk
RD =rigid disk
X =major use
x =minor use

1=γ-Fe_2O_3
2 =γ-Fe_2O_3 +Co
3 =α-Fe
4 =CrO_2
5 =metal alloy film

where M_s is the saturation magnetization of the sample and q is given by

$$q = \sqrt{\frac{M_s}{2H_k}}\,, \tag{12.23}$$

where H_k is the magnetocrystalline anisotropy field. Also, δ is the ferromagnetic exchange length that defines the wall width w given by

Table 12.3 Parameters relating to longitudinal magnetic recording media type (Ullmann's Encyclopedia of Industrial Chemistry, 1989).

magnetic media type	diameter or width (inch)	track pitch [μm]	bit length [μm]
Computer tape IBM 3480	0.5	700	0.65
flexible disk	5.25	260	3.9
flexible disk	3.5	190	3.1
rigid disk IBM 3380 (E)	14	18	2.25
digital audio tape (R-DAT)	0.15	14	0.33

$$w = \pi \frac{\delta}{q} . \tag{12.24}$$

If the angle of the pyramid of the tip is small enough that

$$q \tan\left(\frac{\theta}{2}\right) \frac{\ell}{\delta} << 1 , \tag{12.25}$$

then it is possible to solve for the force derivative acting on the tip whose saturation magnetization is M_t, yielding

$$F_1 = 2w\mu_0 M_s M_t \frac{\tan^2(\theta/2)}{\cosh(qx/\delta)} \zeta . \tag{12.26}$$

Here

$$\zeta = \zeta_1 + \zeta_2 + \zeta_3 + \zeta_4 + \zeta_5 , \tag{12.27}$$

where ζ_i are given by

$$\zeta_1 = (z^2 - w^2) \frac{(2z + \ell)\ell}{[z(z + \ell) + w^2]^2 + w^2\ell^2} , \tag{12.28}$$

$$\zeta_2 = 2z \frac{z + \ell}{(z + \ell)^2 + w^2} , \tag{12.29}$$

$$\zeta_3 = - \frac{2z^2}{z^2 + w^2} , \tag{12.30}$$

$$\zeta_4 = -\ln\left[\frac{z^2 + w^2}{(z + \ell)^2 + w^2}\right], \tag{12.31}$$

and

$$\zeta_5 = -\frac{2z}{w}\,\text{atn}\left[\frac{w\ell}{z(z + \ell) + w^2}\right]. \tag{12.32}$$

12.3.4. Magnetic Lines: Polar Orientation

Hartmann (1990a) developed a model that gives the vertical and horizontal components of the stray magnetic field of a Bloch wall in a magnetic sample shown in Fig. 12.2 (see also, Abraham and McDonald 1990). In this model, which has been used in analyzing Bitter patterns, the field components are given by

$$H_x(x, z) = \frac{1}{2\pi} M_s \int_{-\pi/2}^{\pi/2} \frac{\tan\theta\, d\theta}{\cosh[q\,(x + z\tan\theta)/\delta]} \tag{12.33}$$

and

$$H_z(x, z) = \frac{1}{2\pi} M_s \int_{-\pi/2}^{\pi/2} \frac{d\theta}{\cosh[q\,(x + z\tan\theta)/\delta]}. \tag{12.34}$$

The magnetic field of a tip having an arbitrary geometry can be calculated using a Taylor expansion centered at the apex of the tip and the interaction between the magnetic fields of the tip and the sample gives rise to the force derivative sensed by the tip.

Wadas and Grütter (1989) and Wadas (1989) developed a model that gives the vertical and horizontal components of the stray magnetic field attributable to Bloch line walls on a magnetic sample for a polar orientation shown in Fig. 12.3. The fields are given by

$$H_x(x, z) = 8\,M_s \sum_{n=0}^{\infty} \frac{(-1)^n}{2n + 1} [1 - \exp(-\xi d)] \exp(-\xi z) \sin(\xi x) \tag{12.35}$$

and

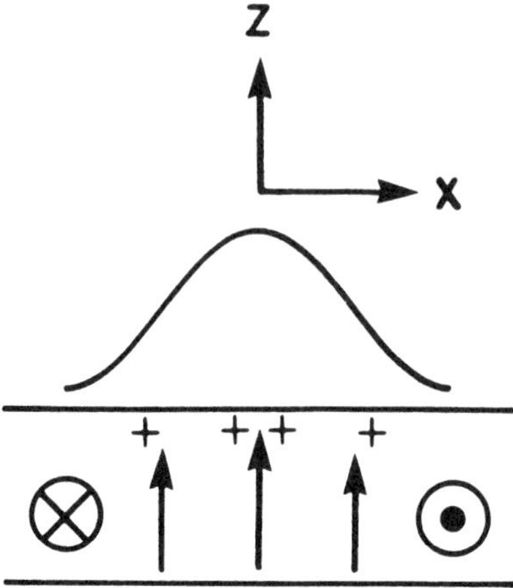

Fig. 12.2 The geometry of the microfield generated by magnetic "charges" (+).

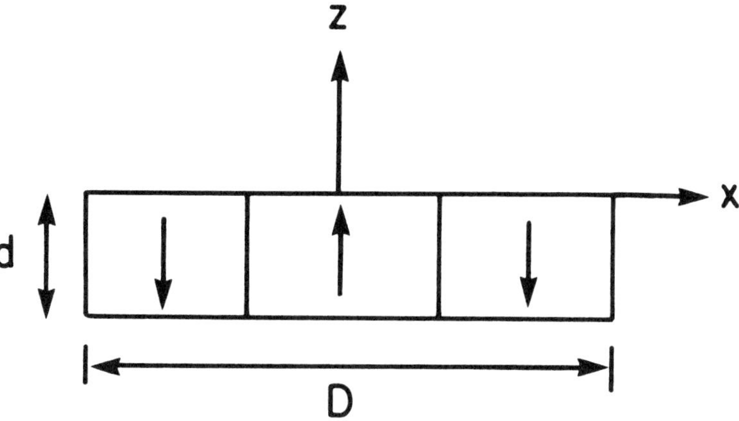

Fig. 12.3 The geometry of the magnetic domains where D is the periodicity and d is the sample height.

$$H_z(x,z) = 8\ M_s \sum_{n=0}^{\infty} \frac{(-1)^n}{2n+1} [1 - \exp(-\xi d)] \exp(-\xi z) \cos(\xi x)\ . \qquad (12.36)$$

They find that the vertical force acting between the tip and the sample is given by

$$F_z(x,z) = \sum_{n=0}^{\infty} \left[A_1 A_2 A_3 \frac{\exp(pz')}{p^2 + r^2} \right]_0^{\ell}, \tag{12.37}$$

taken between 0 and $z' = \ell$. Here the coefficients A_i are given by

$$A_1 = (b + z')\,(p \sin k - r \cos k) - \frac{(p^2 - r^2)\sin k - 2pr \cos k}{p^2 + r^2}, \tag{12.38}$$

$$A_2 = -32\, M_s \frac{(-1)^n}{2n+1} [1 - \exp(-\xi d)] \tan \alpha , \tag{12.39}$$

and

$$A_3 = \exp(-\xi z)\, [M_z \cos(\xi x) + M_x \sin(\xi x)] , \tag{12.40}$$

where $p = -\xi$ and α is half the cone angle of the pyramid,

$$\xi = \frac{2\pi}{D}(2n + 1) , \tag{12.41}$$

$$r = \xi \tan \alpha , \tag{12.42}$$

and

$$k = \xi\, (z' + b) \tan \alpha . \tag{12.43}$$

The sample saturation magnetization M_s in this case points either up or down, while the saturation magnetization of the tip has components in both the x and z directions. D and d are the periodicity and thickness of the domains. The parameters associated with the tip consist of the angle of the pyramid, its total height, $b + \ell$, the length b of the clipped part of the pyramid, and its magnetization components M_x and M_z.

Mansuripur (1989a,b) calculated the tip-sample interactions taking full account of the micromagnetics of the tip. He divided the tip into a large number of small cubes corresponding to crystallites whose magnetic moment depends on the total field acting on each one. The external field H_{eff} consisted of local anisotropy, exchange interaction with nearest neighbors, and the demagnetizing field produced by all the other magnetic moments, as well as the field of the tip stem and the sample. The dynamics of the magnetic moment of each cube is given by the Landau-Lifshitz-Gilbert equation,

$$\frac{\partial \mathbf{m}}{\partial t} = \gamma \mathbf{m} \times \left[\mathbf{H}_{eff} + \frac{\alpha}{\gamma} \frac{1}{|\mathbf{m}|} \frac{\partial \mathbf{m}}{\partial t} \right], \tag{12.44}$$

where γ is the gyromagnetic ratio and α is a viscous damping coefficient. The magnetization of each cube belonging to the tip was calculated self-consistently using about 500 iterations, corresponding to 300 ps, until the interaction energy was properly minimized. Forces were then calculated for the horizontal and perpendicular directions of magnetic domains oriented in the polar direction.

12.3.5. Magnetic Lines: Longitudinal and Polar Orientations

Wadas and Güntherodt (1990) [see also, Wadas et al. (1990a)] developed a model that gives the vertical component of the force acting on the tip attributable to Bloch line walls on a magnetic sample, as shown in Fig. 12.4. Here the magnetic domains have a mixture of polar and longitudinal orientation. The vertical component of the force is given by

$$F_z(x,y) = - \sum_{n=0}^{\infty} A_1 \, A_2 \, A_3 \, A_4 \, \frac{\exp(pz')}{p^2 + r^2} , \tag{12.45}$$

where A_1, A_2, and A_3 are given by Eqs. (12.39) through (12.41),

$$A_4 = \cos\,(\xi w) , \tag{12.46}$$

for geometry (a) and (*b*), and

$$A_4 = \cos\,(\xi w) + \sin\,(\xi w) , \tag{12.47}$$

for geometry (*c*).

12.3.6. Magnetic Lines: Longitudinal Orientation

Rugar et al. (1990) developed a comprehensive model that includes the vertical component of force acting on the tip by the stray magnetic field attributable to Bloch line walls on a magnetic sample where the orientation of the domains is longitudinal, as shown in Fig. 12.4(b). The saturation magnetization of the sample, $\mathbf{M}^s$, is given by

$$\mathbf{M}^s(x) = -\hat{\mathbf{x}} \, \frac{2}{\pi} \, M_r \, \text{atn} \left(\frac{x}{w} \right), \tag{12.48}$$

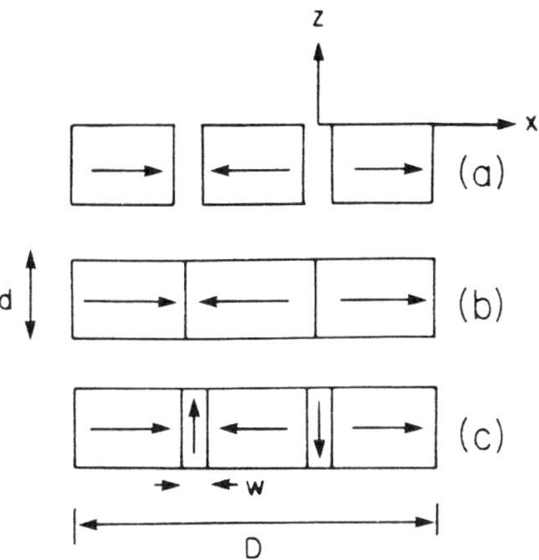

Fig. 12.4 The geometry of the magnetic domains where D is the periodicity of the domain pattern, w is the width of a polar domain, and d is the domain height.

and w characterizes the width of the transition. The horizontal and vertical components of the magnetic field, which are given by

$$H_x(x,z) = 4\ M_r \left[\operatorname{atn}\left[\frac{x(z+d)}{x^2 + w^2 + w(d+z)}\right] - \operatorname{atn}\left[\frac{xz}{x^2 + w^2 + wz}\right]\right] \tag{12.49}$$

and

$$H_z(x,z) = 2\ M_r \log\left[\frac{x^2 + (z + d + w)^2}{x^2 + (w + z)^2}\right], \tag{12.50}$$

are used to calculate the force derivative acting on the tip. The total stray field resulting from the series of magnetic lines is given by

$$\mathbf{H}_{total} = \sum_n (-1)^n\ \mathbf{H}(x - nd, z)\ . \tag{12.51}$$

If the tip is assumed to behave as a uniformly magnetized sphere, whose field is that of a single magnetic dipole, then the force derivative is simply given by

$$F_1(\mathbf{r}) = m_x \frac{\partial^2 H_x}{\partial z^2} + m_y \frac{\partial^2 H_y}{\partial z^2} + m_z \frac{\partial^2 H_z}{\partial z^2}\ . \tag{12.52}$$

If the lever is parallel to the surface of the sample, then the force derivative is given by

$$F_1(\mathbf{r}) = \int_{tip} \sum_{i=x,y,z} M_i^T(\mathbf{r}') \frac{\partial^2 H_i(\mathbf{r} + \mathbf{r}')}{\partial z^2} dV' , \qquad (12.53)$$

where M_i^T are the components of the magnetization of the tip.

Schönenberger and Alvarado (1989, 1990*a*) present a comprehensive treatment of the tip-sample interaction in magnetic force microscopy. They start with a sample having equally spaced longitudinal magnetic charge lines, shown in Fig. 12.5, expressing the stray field (Mamin et al. 1988) in complex notation,

$$H_x + iH_z = M_r \frac{d}{w} \frac{\sin[\pi(x' + iz')]}{|\sin[\pi(x' + iz')]|^2} . \qquad (12.54)$$

Here d is the film thickness, w is the line width, $x' = x/w$, $z' = z/w$, and M_r is the remanent magnetization. Assuming the tip to be a single domain, they expand its field in a two-dimensional Fourier series,

$$\mathbf{H}(\mathbf{k}, z) = \frac{1}{2\pi} \int_0^\infty \exp(-i\mathbf{k}\cdot\mathbf{x})\ \mathbf{H}(\mathbf{x}, z)\ d^2\mathbf{x} . \qquad (12.55)$$

Here the tip has a cone shape with a sharp angle γ and bottom radius R, a domain length L, and volume V. They find the tip-sample force for the three regions

$$\mathbf{F}(\mathbf{k}, d) = \mu_0 M_t\ kV\ \mathbf{H}(\mathbf{k}, d) \qquad kR' \ll 1,\ kL \ll 1 , \qquad (12.56)$$

$$\mathbf{F}(\mathbf{k}, d) = \mu_0 M_t\ 2\pi \frac{c^2}{k^2}\ \mathbf{H}(\mathbf{k}, d) \qquad kR' \ll 1 \ll kL , \qquad (12.57)$$

and

$$\mathbf{F}(\mathbf{k}, d) = \mu_0 M_t\ \pi R^2 \exp(-k^2 R^2/4)\ \mathbf{H}(\mathbf{k}, d) \qquad kR' \gg 1 . \qquad (12.58)$$

Here $\mathbf{x} = (x_1, x_2)$, $\mathbf{k} = (k_1, k_2)$, $x_1 = x$, $x_2 = y$, $c = \tan(\gamma/2)$, and $R' = R/c$. Also, for sharp domain transitions across the sample, H is given by

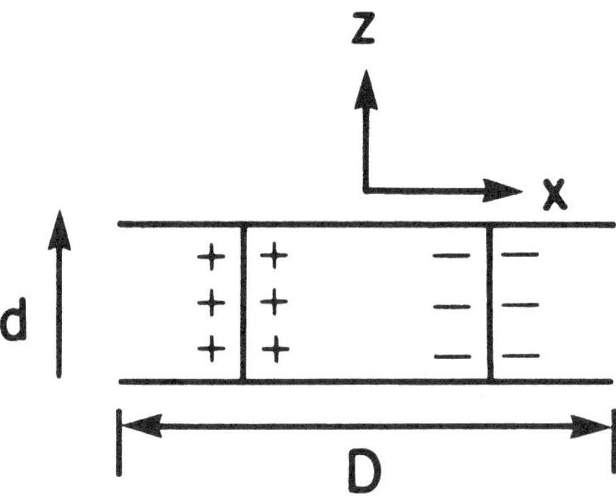

Fig. 12.5 The geometry of the domains generated by equally spaced magnetic "charge" lines, where D is the periodicity and d is the domain height.

$$H = \frac{2}{\pi} M_r \; kd \; \exp(-kd) \; . \tag{12.59}$$

A detailed treatment of these results leads to the understanding of why the tip domain length determines the transition from a single pole to a dipole behavior, and why magnetic images sometimes resemble stray field distributions rather than their derivative.

12.4. Principles of Operation

In this section we develop the theory of operation of magnetic force microscopy for systems using a bimorph-driven lever, a sample-driven lever, and a voltage-driven lever. We derive general expressions that give the change in amplitude of vibration of the lever supporting the force-sensing conducting tip in terms of the parameters of the system and discuss the signal-to-noise ratio. The amplitude of vibration will be a measure of the magnetic properties of the sample, the topography, the dielectric constant ϵ, and the charge or the potential on the surface of the conducting sample that can be covered by a dielectric film. We can then raster scan the tip across the sample using any of the systems described in Chapters 4 through 10 and obtain a mapping of these properties. The electrostatic contributions have been discussed in detail in Chapter 11, where we have chosen to use, for simplicity, the model of a plane-parallel capacitor for the tip-sample interaction rather than the more accurate sphere-plane geometry. We note that while the electric component of the force is always attractive, the magnetic component can be either attractive or repulsive. We also note that in Chapter 11 the externally applied potential and the potential of the charge distribution could in principle exchange energy. The energy, therefore, is composed of the square of the sum of the potentials. In contrast, the electrostatic and magnetostatic potentials in this section are independent, and the total energy is composed of the sum of their squares.

12.4.1. Bimorph-Driven Lever

Figure 12.6 shows a schematic of a typical system where the lever is mounted on a bimorph vibrating with amplitude a and frequency ω. In this configuration we get

$$u = u_0 + a \exp(i\omega t) , \tag{12.60}$$

$$z = z_0 + \zeta , \tag{12.61}$$

$$\zeta = A_b(\omega) \exp[i(\omega t - \theta)] , \tag{12.62}$$

and

$$g = 0 . \tag{12.63}$$

Here $A_b(\omega)$ and ω are the amplitude and frequency of vibration of the lever and θ is a phase angle. The equation of motion for this system is

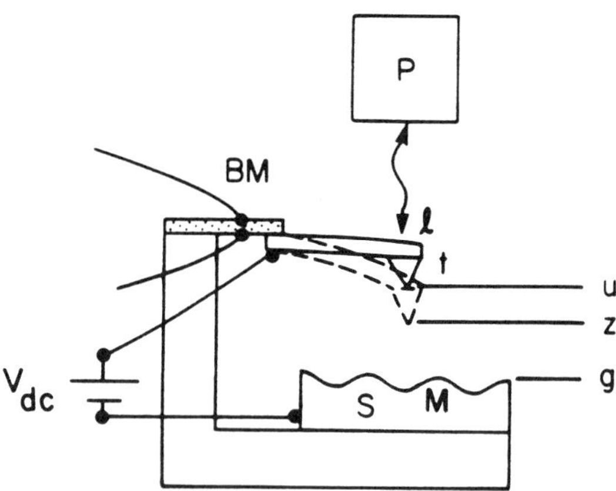

Fig. 12.6 Bimorph-driven lever with magnetic sample and applied potentials, where M denotes a magnetic sample (S).

$$m \frac{\partial^2 z}{\partial t^2} + \gamma \frac{\partial z}{\partial t} + k(z - u) = F(z) \,. \tag{12.64}$$

We assume now that the electrostatic and magnetostatic forces have only dc components and that no charge distribution exists on the surface of the sample. The forces with which we are concerned have derivatives in the direction of vibration of the lever and can therefore be written as

$$F = F_0 + F_1 \zeta \,, \tag{12.65}$$

where the force F and its derivative F_1 are given by

$$F(z) = F_m - \epsilon \frac{SV^2{}_{dc}}{2z^2} \tag{12.66}$$

and

$$F_1 = F_{1,m} + \epsilon \frac{SV^2{}_{dc}}{z^3{}_0} \,, \tag{12.67}$$

respectively, and m denotes the magnetic component. Since we are not interested in the exact electrostatic force, we approximate the tip-sample capacitance by that of a parallel-plate capacitor. We have also assumed that $V = V_{dc}$ and that no electric charges exist on the magnetic sample. The average position of the lever is given by

$$k(z_0 - u_0) = -F_m + \epsilon \frac{SV^2{}_{dc}}{2z^2{}_0} \,. \tag{12.68}$$

The effective spring constant,

$$k' = k - F_{1,m} - \epsilon \frac{SV^2{}_{dc}}{z^3} \,, \tag{12.69}$$

will shift the resonance frequency of the lever to

$$\omega'_0 = \sqrt{\frac{k'}{m}} \,. \tag{12.70}$$

In analogy to Eq. (11.58), we get the variation of the amplitude of vibration for operation at $\omega = \omega_m$ on the slope of the resonance curve, in terms of the variation in the force derivative,

$$\delta A_b(\omega = \omega_m) = \frac{2}{3\sqrt{3}} \frac{Q}{k} A_b \ \delta\left[F_{1,m} + \epsilon \frac{SV^2{}_{dc}}{z_0{}^3}\right] . \tag{12.71}$$

12.4.2. Sample-Driven Lever

Figure 12.7 is a schematic of a typical system where the sample is mounted on a bimorph vibrating with amplitude a and frequency ω. In this configuration we get

$$u = u_0 , \tag{12.72}$$

$$z = z_0 + \zeta , \tag{12.73}$$

$$\zeta = A_s(\omega) \exp[i(\omega t - \theta)] , \tag{12.74}$$

and

$$g = g_0 + a \exp(i\omega t) . \tag{12.75}$$

Here $A_s(\omega)$ and ω are the amplitude and frequency of vibration of the lever and θ is a phase angle. The equation of motion for this system is

$$m \frac{\partial^2 z}{\partial t^2} + \gamma \frac{\partial z}{\partial t} + k(z - u) = F(z-g) , \tag{12.76}$$

where F and F_1 are given by

$$F(z) = F_m - \epsilon \frac{SV^2{}_{dc}}{2(z-g)^2} \tag{12.77}$$

and

$$F_1 = F_{1,m} + \epsilon \frac{SV^2{}_{dc}}{(z-g)^3} . \tag{12.78}$$

The average position of the lever is determined by equating the electrostatic and magnetostatic forces to the restoring force,

$$k(z_0 - u_0) = -F_m + \epsilon \frac{SV^2{}_{dc}}{2(z_0 - g_0)^2} . \tag{12.79}$$

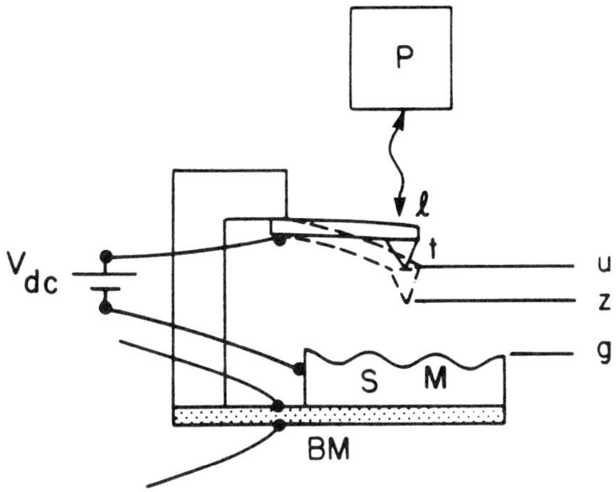

Fig. 12.7 Sample-driven lever with magnetic sample and applied potentials.

The effective spring constant in this case is

$$k' = k - F_{1,m} - \epsilon \frac{SV^2_{dc}}{(z_0 - g_0)^3} . \tag{12.80}$$

In analogy with Eq. (11.66), we get for the amplitude of vibration on resonance, $A_s(\omega = \omega'_0)$,

$$A_s(\omega = \omega_0') = \frac{aQ}{k} [F_{1,m} - \epsilon SV^2(z - g)^2)] . \tag{12.81}$$

As shown by Eq. (11.70), if $k > QF_1$, it is advantageous to operate on resonance where $\omega = \omega'_0$.

12.4.3. Voltage-Driven Lever

Figure 12.8 shows a schematic of a system where the lever is driven by applying a dc magnetic field and an ac voltage V_{ac} with frequency Ω and amplitude V between the tip and the sample. In this configuration we have for u, z, ζ, and g,

$$u = u_0 , \tag{12.82}$$

$$z = z_0 + \zeta , \tag{12.83}$$

$$\zeta = A_v(\omega) \exp[i(\omega t - \theta)] , \tag{12.84}$$

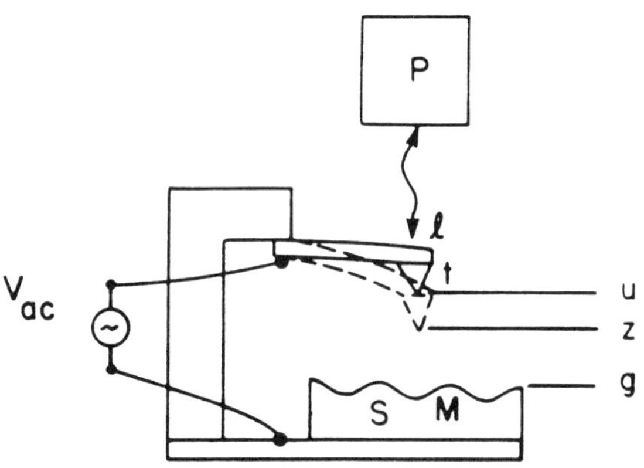

Fig. 12.8 Voltage-driven lever with magnetic sample and applied potentials.

and

$$g = 0 \, , \tag{12.85}$$

where A_V and ω are the amplitude and frequency of vibration of the lever. To determine the external force acting on the lever, we assume a driving voltage V given by

$$V_{ac} = V \sin(\Omega t) \, . \tag{12.86}$$

The resulting forces and their derivatives are

$$F_{dc} = F_m + \frac{1}{4} C' V^2 \, , \tag{12.87}$$

$$F_{1,dc} = F_{1,m} + \frac{1}{4} C'' V^2 \, , \tag{12.88}$$

$$F_{2\Omega} = - \frac{1}{4} C' V^2 \, , \tag{12.89}$$

and

$$F_{12\Omega} = - \frac{1}{4} C'' V^2 \, , \tag{12.90}$$

where

$$C' = -\frac{\epsilon S}{(z-g)^2} \tag{12.91}$$

and

$$C'' = \frac{2\epsilon S}{(z-g)^3} . \tag{12.92}$$

The equation of motion for this system, to first order, is

$$m\frac{\partial^2 z}{\partial t^2} + \frac{\omega_0 m}{Q}\frac{\partial z}{\partial t} + k(z-u) = F_{dc} + F_{2\Omega} + F_{1dc}\,\zeta . \tag{12.93}$$

The average position of the lever will be determined by equating the restoring forces to the electrostatic and magnetostatic forces, yielding

$$k(z_0 - u_0) = F_m + \frac{1}{2}C'V^2 . \tag{12.94}$$

Now, however, the lever can vibrate only at $\omega = 2\Omega$ with

$$\zeta = A_{v}(\omega = 2\Omega) . \tag{12.95}$$

The equation of motion yields the amplitude of vibration

$$A_{v}(\omega = 2\Omega) = \frac{|F_{2\Omega}|}{k'}\frac{Q}{[Q^2[1-\omega^2/\omega_0'^2]^2 + \omega^2\omega_0^2/\omega_0'^4]^{1/2}} . \tag{12.96}$$

Operating at $\omega = \omega_m$ gives

$$\delta A_{v}(\omega = 2\Omega = \omega_m) = \frac{2}{3\sqrt{3}}\frac{Q}{k}A_{v}\,\delta\left[F_{1,m} + \frac{1}{2}C'V^2\right] . \tag{12.97}$$

12.5. Noise Considerations

We found that the thermal noise on resonance of the lever is given by

$$\langle\delta z^2\rangle_\ell = \frac{4KTBQ}{k\omega_0} . \tag{12.98}$$

We will now apply this result to our three cases and find the minimum value of the force derivative δF_{1dc} that is limited by lever noise.

12.5.1. Bimorph-Driven Lever

From Eq. (12.71) we get for operation at ω_m,

$$\delta F_{1,m} = \frac{3\sqrt{3}}{2} \frac{k}{aQ^2} \sqrt{\frac{4KTBQ}{k\omega_0}} . \tag{12.99}$$

12.5.2. Sample-Driven Lever

Here we find for operation at ω_0, using Eq. (12.81),

$$\delta F_{1,m} = \frac{k}{aQ} \sqrt{\frac{4KTBQ}{k\omega_0}} . \tag{12.100}$$

12.5.3. Voltage-Driven Lever

Here we have for operation at ω_m, using Eq. (12.97),

$$\delta F_{1,m} = \frac{3\sqrt{3}\, k}{2QA_{2\Omega}} \sqrt{\frac{4KTBQ}{k\omega_0}} . \tag{12.101}$$

12.6. Applications

The purpose of magnetic force microscopy is to measure magnetic interactions, for which we use the voltage-driven lever system as an example and operate at $\omega = 2\Omega = \omega_m$, basing the analysis on Eq. (12.71).

12.6.1. Modeling the Magnetic Interaction

To measure the magnetic force we use

$$\delta A_v(\omega = 2\Omega = \omega_m) = \frac{2}{3\sqrt{3}} \frac{Q}{k} A_v \, \delta\left[F_{1,m} + \frac{1}{2} C'V^2\right] . \tag{12.102}$$

12.7. Performance

Several magnetic materials have been investigated using scanning force microscopy. Table 12.4 is a summary of these materials, part of which were used for theoretical discussions, part were imaged with magnetic force microscopy, and some were used for both (Figs. 12.9 through 12.14).

Table 12.4 Summary of magnetic materials probed by magnetic force microscopy.

Material	Reference
Co	Wadas (1988a)
Co alloy	Abraham et al. (1988a,b)
Co-alloy on Si	Mamin et al. (1988)
CoNi	Grütter et al. (1988)
CoNi	Hosaka et al. (1990)
Co-alloy disk	Rugar et al. (1989)
CoPtCr	Rugar et al. (1990)
CoPtCr	Albrecht et al. (1990a)
CoSm	Rugar et al. (1990)
CoCr on NiFe	Rugar et al. (1990)
CoCr	Abraham et al. (1988b)
CoCr	den Boef (1990)
CoCr	Grütter et al. (1989, 1990a)
CoCrTa	Moreland and Rice (1990b)
iron	Allenspach et al. (1987)
iron	Wadas (1988b)
iron	Göddenhenrich et al. (1988)
iron	Hartmann and Heiden (1988)
iron	Hartmann (1990)
γ-FeO_3	Rugar et al. (1990)
Ni	Saenz et al. (1987)
Ni	Grütter et al. (1988)
NiFe	Mamin et al. (1989)
FeNdB	Grütter et al. (1988, 1990a,b)
TbFe	Saenz et al. (1988)
TbFe	Martin et al. (1988a)
TbFe	Abraham et al. (1988a,b)
TbFe	Wadas and Grütter (1989)
TbFe	Wadas (1989)

Table 12.4 (continued) Summary of magnetic materials probed by magnetic force microscopy.

Material	Reference
TbFeCo	den Boef (1990)
TbFeCo	Rugar et al. (1988)
TbFeCo	Hobbs et al. (1989)
TbFeCo	Wickramasinghe (1990)
TbFeCo	Sarid et al. (1988, 1990)
IBM 3380E thin film recording head	Martin and Wickramasinghe (1987)
IBM 337X thin film recording head	Schönenberger et al. (1990)
Superconductor	Hug et al. (1990)
Tape	Iijima and Yasuda (1988)

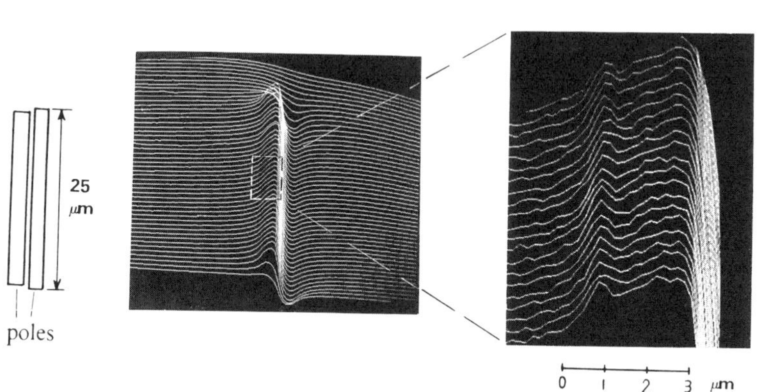

Fig. 12.9 Magnetic image of an IBM 3380 recording head (Courtesy: M. P. O'Boyle, D. W. Abraham, and K. Wickramasinghe, IBM).

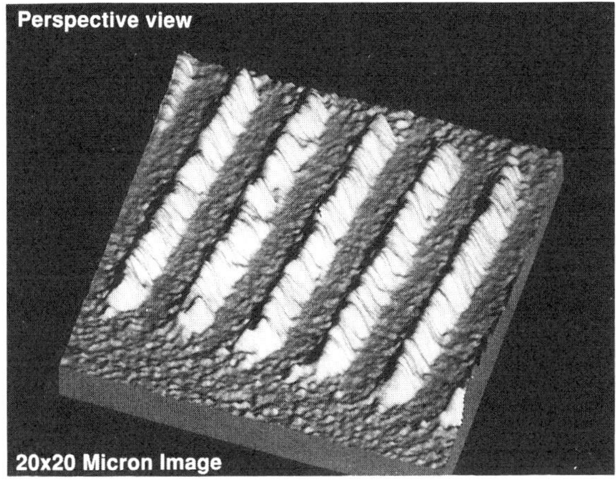

Fig. 12.10 Magnetic image of a bit track in thin film CoPtCr disk (Courtesy: Y. Martin and K. Wickramasinghe, IBM).

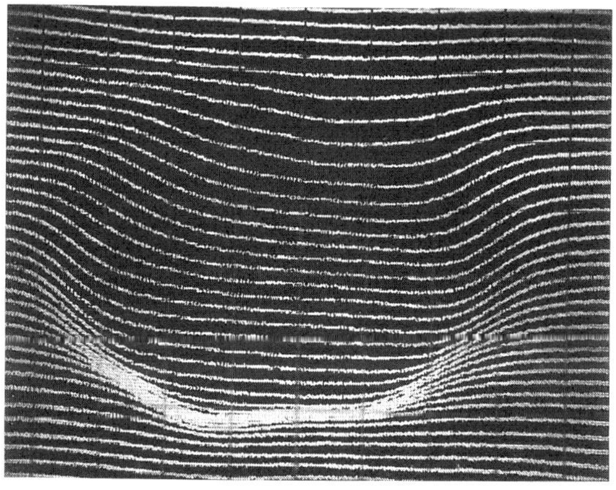

0.5 μm

Fig. 12.11 Magnetic image of a TbFeCo domain used for magneto-optic storage media (D. Iams and D. Sarid).

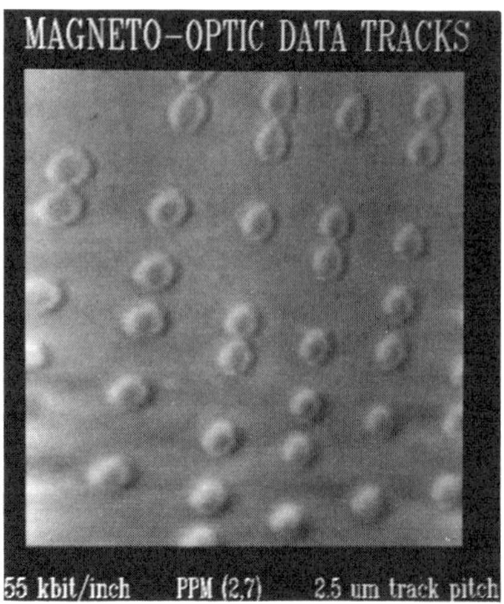

Fig. 12.12 Magnetic image of data tracks (running vertically) on a magneto-optical TbFeCo disk. The bits have $\simeq 0.8\ \mu m$ diameter and the track pitch is $2.5\ \mu m$ (Courtesy: P. Grütter and D. Rugar, IBM).

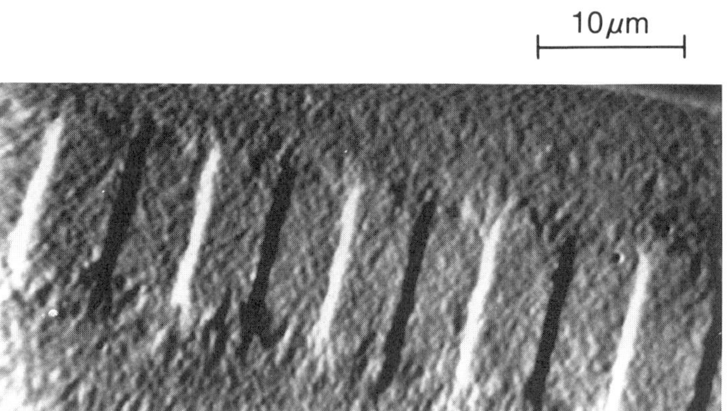

Fig. 12.13 Magnetic image of a bit track in Co-alloy disk (Courtesy: H. J. Mamin and D. Rugar, IBM).

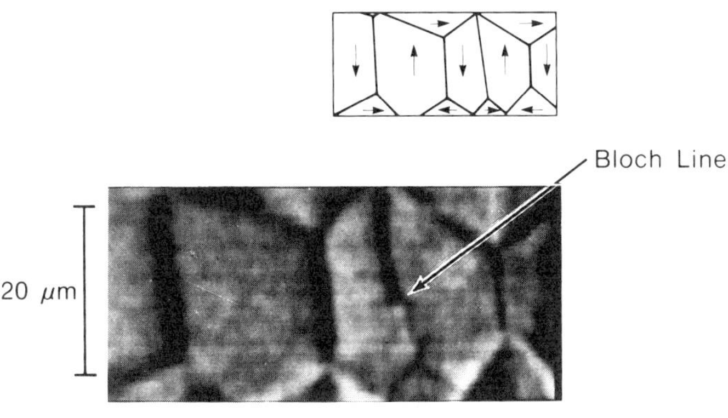

Fig. 12.14 Magnetic image of domains in Permalloy film where the insert shows the orientation of the domains (Courtesy: H. J. Mamin and D. Rugar, IBM).

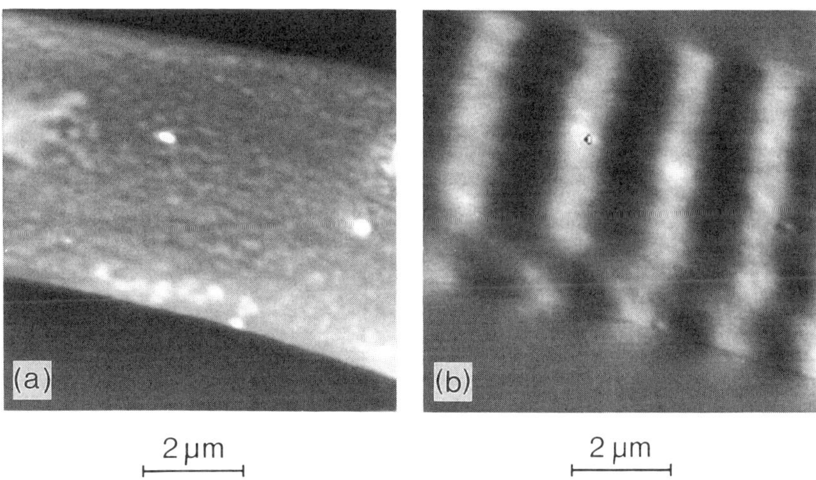

Fig. 12.15 Topographic and magnetic images of sputtered CoPt films obtained simultaneously (Courtesy: C. Schönenberger and S. F. Alvarado, IBM).

12.8. Summary

We found that the relative changes in the amplitude of vibration in terms of force derivatives for our three methods of operation are:

Bimorph-driven lever at $\omega = \omega_m$,

$$\delta A_b(\omega = \omega_m) = \frac{2}{3\sqrt{3}} \frac{Q}{k} A_b \, \delta\left[F_{1,m} + \epsilon \frac{SV^2{}_{dc}}{z_0{}^3}\right].$$

Sample-driven lever at $\omega = \omega'_0$,

$$A_s(\omega = \omega_0') = \frac{aQ}{k} [F_{1m} - \epsilon SV^2(z - g)^2)] .$$

Voltage-driven lever at $\omega = 2\Omega = \omega_m$,

$$\delta A_v(\omega = 2\Omega = \omega_m) = \frac{2}{3\sqrt{3}} \frac{Q}{k} A_v \, \delta\left[F_{1,m} + \frac{1}{2} C'V^2\right].$$

13
Atomic Force Microscopy

13.1. Introduction

In this chapter we review the microscopic and macroscopic theories that model interatomic (as well as intermolecular) forces, and present theoretical applications for, and experimental results of, atomic force microscopy. Interatomic forces act between any two atoms or molecules and can be classified according to whether they are repulsive or attractive, short or long range. In equilibrium, typical interatomic spacings are on the order of 2 to 3 Å, below which interatomic forces are always repulsive and short range. At larger atomic separations, however, the long-range interatomic forces can be either repulsive or attractive. The theory involving the short-range repulsive forces is the more difficult one, because at close proximity, two atoms sense mainly each other, and the interaction is atom specific. In contrast, when the distances involved are large relative to atomic distances, only averages of atomic properties must be taken into account. At large distances, therefore, one can model the interaction as if it were between two macroscopic bodies, and the theory is simpler. A significant simplification in the calculation of short-range interatomic forces can be obtained by realizing that (through the Hellman-Feynman theorem) once the distribution of electronic charges is known for each of the molecules separately, the interaction between any two can be calculated using classical electromagnetic theory. One can, therefore, come up with a simplified picture of electronic charge distributions and, in quite a straightforward manner, deduce the main features of interatomic interactions using classical methods.

This chapter presents basic microscopic and macroscopic theories following the treatment by Israelachvili (1985), theories of tip-sample interactions, principles of operation, and a summary of performance. In the section on basic microscopic theory, we treat physical bonds, Coulomb forces, ionic forces, ion-dipole interaction, dipole-dipole interaction, electronic polarizability, molecular-orientation polarizability, dipole-induced dipole interaction, van der Waals dispersion forces, the McLachlan model, and the Lennard-Jones potential. The section on basic macroscopic theory gives examples of macroscopic interactions, the Derjaguin approximation, the Lifshitz model, and surface energy. We then survey theories of tip-sample interactions, starting with early theories and continuing with the modern theories, using molecular-dynamics simulations. The latter include semiconductor-tip and semiconductor-sample, metal-tip and semiconductor-sample, and metal-tip and metal-sample interactions. We then describe the principles involved in the operation of an atomic force microscope that maps normal and lateral forces across a surface, and the dynamics of the lever-tip-sample approach. We conclude the theoretical discussion with a comparison of atomic, electrostatic, and magnetostatic interactions. The final section presents a summary of performance where images of crystals, structures, biomaterials, and polymers have been obtained, and processes, modification, indentation, and tribology characterized. For a general discussion, see for example Meyer et al. (1992).

13.2. Intermolecular Microscopic Interactions

In developing a theory for interatomic forces, it is useful to consider interactions that are proportional to a given power law, or a combination of power-law terms. In general, we can write the attractive interaction between two microscopic or macroscopic bodies, for $r > \sigma$, as

$$w(r) = -C/r^n \ , \tag{13.1}$$

where n is an integer and $w(r<\sigma) = \infty$. As we shall see later, this assumption is correct for distances larger than, say, 10 Å. To find the contribution to this interaction from all molecules that surround and act on a given molecule, we consider the number density $\rho(r)$ of these molecules and integrate the interaction on the whole volume. For the total interaction energy, therefore, we get

$$w = \int_{\sigma}^{\ell} w(r)\ \rho(r)\ 4\pi r^2 dr \ , \tag{13.2}$$

where σ and ℓ are the radii of the given molecule and of the total system, respectively. In the case were $\sigma << \ell$, which is realized when the given molecule interacts with a much larger body, we get

$$w = -\frac{4\pi C\rho}{(n-3)\sigma^{n-3}} \ . \tag{13.3}$$

Now, it has been observed experimentally that the interaction between a molecule and the whole system surrounding it, being finite, is relatively short-ranged, extending to distances of only several hundred angstroms.

Consequently, the coefficient n in the exponent of Eq. (13.1) must fulfill the requirement that

$$n > 3 \ . \qquad (13.4)$$

We find, therefore, that the interaction is mainly between closest neighbors, and does not depend on the macroscopic size of the system. The interaction energy between two molecules, being distance-dependent, has a field of force associated with it. This is called the free energy because it is actually the available mechanical energy. An important consideration, when assessing the strength of a given interaction between two molecules, is the comparison of its energy with the thermal energy, KT, where T is the temperature of the thermal bath surrounding the molecules. If $|w| \ll$ KT, the thermal bath will dominate, and forces associated with this interaction will be "smeared out." In contrast, if $|w| \gg$ KT, the thermal bath will not affect the action of the force, and the two molecules can remain attached to each other for a long period of time. We shall now present a survey of several of the dominant interactions that affect the force-sensing tip and sample of an atomic force microscope.

13.2.1. Covalent Bonds

Covalent or chemical bonds exist between two or more molecules when electrons are shared between them. The number of such bonds for a given atom is its stoichiometry or valency. These bonds are characterized by their strong directionality, which controls the relative orientation of the atoms or molecules within their environment. The range of the forces associated with these bonds is small, on the order of 1 to 2 Å, and the energy per bond is on the order of 100 to 300 KT. Examples of typical covalent bonds are those of H_2O, CH_4, C_2H_4.

13.2.2. Physical Bonds

Physical bonds give rise to forces between unbonded discrete atoms or molecules. Although these forces may be as strong as those of covalent bonds, the electronic clouds of each individual atom are only slightly perturbed by this interaction. Physical bonds do not have a directionality or stoichiometry, are long-range, and participate in interactions that do not involve chemical reactions.

13.2.3. Coulomb Forces

Coulomb forces exist between any two distributions of charges, and for two isolated charges, q_1 and q_2, exhibit an interaction given by

$$w(r) = \frac{q_1 q_2}{4\pi\epsilon} \frac{1}{r} . \tag{13.5}$$

For a typical interatomic distance of r = 2.76 Å, with $q_i = e$ (where e is the electronic charge), the energy is 8.4×10^{-19} J, or approximately 200 KT at room temperature. The two charges must be separated by 560 Å for the energy to equal that of the thermal bath, namely, w = KT, indicating that the interaction is a very strong one indeed. Although Coulomb energies between two isolated charges are inversely proportional to the distance r, their range in a medium filled with other charges is much shorter because these charges tend to screen the interaction, which, therefore, falls off much faster.

13.2.4. Ionic Forces

In ionic crystals, an imbedded charged atom interacts with all of its neighbors. In *a* Na^+Cl^- crystal, for example, the interaction energy will be given by the sum

$$w(r) = -\frac{e^2}{4\pi\epsilon} \frac{1}{r} \left[6 - \frac{12}{\sqrt{2}} + \frac{8}{\sqrt{3}} - \frac{6}{2} + .. \right] , \tag{13.6}$$

which equals

$$w(r) = -1.748 \frac{e^2}{4\pi\epsilon} \frac{1}{r} . \tag{13.7}$$

For r = 2.76 Å, for example, $w = -1.46\times10^{-18}$ J = 352 KT. Here, 1.748 is the Madelung constant, which, although differing from one molecule to the other, has a typical value of approximately 1.6 for monovalent-monovalent crystals, and 5 for monovalent-divalent crystals. The self energy associated with assembling an ion is its Born energy. Assembling a collection of charges from infinity to a given volume requires the energy

$$w(r) = \frac{1}{4\pi\epsilon} \int_0^q \frac{q\,dq}{r} = \frac{q^2}{8\pi\epsilon} \frac{1}{r} , \tag{13.8}$$

where q is the total charge of the ion. Most molecules possess an electric dipole moment $\mu = q\ell$, given in terms of the charges, $\pm q$, and their separation, ℓ. The unit of the dipole moment is a Debye, D = 3.336×10^{-30}, having units of coulomb-meter. Two electronic charges, for example, separated by 1 Å, have a dipole moment equal to 4.8 D, and many molecules that contain covalent bonds have dipole moments on that order. We can write Eq. (13.8) in terms of the dipole moment:

$$w(\ell) = \frac{\mu^2}{4\pi\epsilon}\frac{1}{\ell^3}\,. \tag{13.9}$$

If the ion is moved from one environment to another, with dielectric constants ϵ_1 and ϵ_2, respectively, the change in the Born energy upon this transfer will be given by

$$\Delta w = -\frac{q^2}{4\pi\ell}\left[\frac{1}{\epsilon_1} - \frac{1}{\epsilon_2}\right]. \tag{13.10}$$

13.2.5. Ion-Dipole Interactions

The interaction between an ion and a dipole is given by

$$w(r) = \frac{q\mu\cos\theta}{4\pi\epsilon}\frac{1}{r^2}\,, \tag{13.11}$$

where q is the charge of the ion and θ is the angle between the direction of the dipole moment and the ion. For an angle-averaged interaction, such that KT > $q\mu/(4\pi\epsilon r^2)$, we get

$$w(r) = -\frac{q^2\mu^2}{6(4\pi\epsilon)^2 KT}\frac{1}{r^4}\,. \tag{13.12}$$

13.2.6. Dipole-Dipole Interactions

Dipole-dipole interactions are given by

$$w(r) = \frac{-\mu_1\mu_2}{4\pi\epsilon r^3}\left[2\cos\theta_1\cos\theta_2 - \sin\theta_1\sin\theta_2\cos\phi\right], \tag{13.13}$$

and for an angle-averaged interaction, such that KT > $\mu_1\mu_2/4\pi\epsilon r^3$, the energy is given by

$$w(r) = -\frac{\mu_1{}^2\mu_2{}^2}{3(4\pi\epsilon)^2KT}\,\frac{1}{r^6}\,. \tag{13.14}$$

13.2.7. Electronic Polarizability

An electric field, **E**, will give rise to an induced dipole moment,

$$\boldsymbol{\mu}_{ind} = e\boldsymbol{\ell}\,, \tag{13.15}$$

and a force, $\mathbf{F} = e\mathbf{E}$, which is balanced by the electron-nucleus Coulomb interaction, $e^2\sin\theta/4\pi\epsilon r^2$. Because $\sin\theta = \ell/r$, where r and ℓ are the radius of the electronic orbit and the component of the orbit in the direction of the field, respectively, we obtain the electronic dipole moment

$$\boldsymbol{\mu}_{ind} = 4\pi\epsilon r^3\mathbf{E}\,. \tag{13.16}$$

The polarizability,

$$\alpha_0 = \frac{\mu_{ind}}{E}\,, \tag{13.17}$$

is therefore given by

$$\alpha_0 = 4\pi\epsilon r^3\,. \tag{13.18}$$

13.2.8. Molecular Orientational Polarizability

A dipolar molecule with a permanent dipole moment, $\boldsymbol{\mu}$, will orient itself under the influence of an external electric field. Taking into account the thermal bath that will try to disorient the molecule, we obtain for an average induced moment the orientational polarizability

$$\alpha = \frac{\mu^2}{3KT}\,. \tag{13.19}$$

The total polarizability is therefore given by

$$\alpha = 4\pi\epsilon r^3 + \frac{\mu^2}{3KT}\,, \tag{13.20}$$

which is the Debye-Langevin equation.

13.2.9. Dipole-Induced Dipole Interactions

Dipole-induced dipole interactions can be calculated from the electric field of a dipole, which is given by

$$E = \frac{\mu(1 + 3\cos^2\theta)^{1/2}}{4\pi\epsilon} \frac{1}{r^3} . \tag{13.21}$$

Because $w = -\alpha_0 E^2/2$, where α_0 is the polarizability, we get

$$w(r) = -\frac{\mu^2\alpha_0(1 + 3\cos^2\theta)}{2(4\pi\epsilon)^2} \frac{1}{r^6} . \tag{13.22}$$

By angle averaging $\langle\cos^2\theta\rangle = 1/3$, we obtain

$$w(r) = -\frac{\mu^2\alpha_0}{(4\pi\epsilon)^2} \frac{1}{r^6} . \tag{13.23}$$

for the interaction. If two atoms have different dipole moments and polarizabilities, μ_i and α_{0i}, respectively, then Eq. (13.23) becomes

$$w(r) = -\frac{\mu_1{}^2\alpha_{02} + \mu_2{}^2\alpha_{01}}{(4\pi\epsilon)^2} \frac{1}{r^6} . \tag{13.24}$$

13.2.10. van der Waals Dispersion Forces

van der Waals dispersion forces act between any given number of atoms, including neutral ones such as He and CO_2. The term *dispersion* applies here because these forces are associated with the dispersion of light in matter. van der Waals dispersion forces, which can be either attractive or repulsive and have a range that extends from 2 Å to more than 1000 Å, must be calculated self-consistently. In other words, the interaction between two molecules is affected by the introduction of a third one, and one cannot simply add the individual pair potentials to obtain the total interaction. The origin of van der Waals dispersion forces can be understood by considering that, in every atom, there is an instantaneous dipole moment that will always act on a nearby atom to generate an instantaneous induced dipole moment. Instantaneous here means that, at any point in time, the electron can be perceived as being somewhere around its orbit, and only average charge distributions of electronic clouds can, therefore, be spherical. It should be noted, however, that this interaction has a finite speed of propagation and, consequently, retardation effects must be taken into account. It was established both theoretically and experimentally (Israelachvili 1972) that, below 10 Å, retardation effects can be neglected, while for distances larger than 100 Å, they play a major role. For example, it was found (Israelachvili 1972) that for crossed cylinders of mica, the van der Waals interaction is nonretarded in the range of 20 Å to 120 Å, and obeys the $1/z^2$ law. In the 120 Å to 500 Å range, retardation effects arise, and the power law changes to $1/z^{2.9}$. A simple model for van der Waals dispersion forces can be obtained by considering an electron orbiting a proton and equating the Coulomb interaction to $2h\nu$, where $\nu = 3.3\times10^{15}$ Hz is the orbiting frequency of the electron. One obtains for the first Bohr radius, a_0, the smallest orbit possible,

$$a_0 = \frac{e^2}{8\pi\epsilon h\nu} = 0.53\ \text{Å}\ . \tag{13.25}$$

The energy needed to ionize an atom, called the first ionization potential, is equal to $h\nu = 2.2\times10^{-18}$ J. The Bohr atom has an instantaneous dipole moment, which will interact with the polarizability of a second atom, $\alpha_0 = 4\pi\epsilon a^3{}_0$, giving rise to an energy

$$w(r) = -\frac{\alpha^2{}_0 h\nu}{(4\pi\epsilon)^2}\frac{1}{r^6}\ . \tag{13.26}$$

The London result, based on a more accurate calculation, adds a factor of 3/4 to Eq. (13.26). For dissimilar atoms, Eq. (13.26) can be written as

$$w(r) = -\frac{3}{2}\frac{\alpha_{01}\alpha_{02}}{(4\pi\epsilon)^2}\frac{h\nu_1 h\nu_2}{h(\nu_1+\nu_2)}\frac{1}{r^6}\ . \tag{13.27}$$

13.2.11. The McLachlan Model

McLachlan generalized the van der Waals theory of two atoms interacting through a medium, obtaining

$$w(r) = -\frac{6\mathrm{KT}}{(4\pi\epsilon_0)^2}\sum_n \frac{\alpha_1(i\omega_n)\alpha_2(i\omega_n)}{\epsilon^2{}_3(i\omega_n)}\frac{1}{r^6}\ , \tag{13.28}$$

where

$$\hbar\omega_n = \mathrm{KT}\ . \tag{13.29}$$

Equation (13.28) actually includes, by means of the excess polarizabilities, all three contributions to the van der Waals interaction, namely the induction, orientation, and dispersion components.

13.2.12. Lennard-Jones Potential

The main forces involved in interatomic interactions arise from: (1) electrostatic or Coulomb interactions between charges or charge distributions, such as monopoles, dipoles, quadrupoles, and combinations of these interactions; (2) polarization forces, where a distribution of charges in one molecule creates a dipole moment in an adjacent molecule; and (3) quantum-mechanical forces, which give rise to covalent bonding and to repulsive exchange interactions. These last forces derive from Pauli's exclusion principle, which prevents electrons belonging to different molecules from having the same quantum numbers. Consequently, the interaction raises the electrons to higher energy levels, resulting in a repulsive force. It is found experimentally that these repulsive forces grow very steeply as the interatomic distance shrinks. On separating the two atoms, the repulsive force falls to zero at distances on the order of 3 Å, upon which the force becomes attractive. One can model the repulsive potential by either a power-law potential,

$$w(r) = (\sigma/r)^n \ , \tag{13.30}$$

or by an exponential potential,

$$w(r) = c\ \exp(-r/\sigma_0) \ , \tag{13.31}$$

each of which makes the two-atom system compressible. The phenomenological Lennard-Jones potential combines the attractive van der Waals and repulsive atomic potentials, giving

$$w(r) = 4w_0[(\sigma/r)^{12} - (\sigma/r)^6] \ . \tag{13.32}$$

In Eq. (13.32) the energy obtains its minimum value of $w = -w_0$ at a distance $r = 1.12\sigma$, and is zero for a distance $r = \sigma$. The forces are obtained from Eq. (13.32), yielding

$$\mathbf{F}(r) = -24w_0[\sigma^6/r^7 - 2\sigma^{12}/r^{13}]\ \hat{r} \ . \tag{13.33}$$

13.3. Intermolecular Macroscopic Interactions

In this section we use the results of the basic microscopic theory to obtain the interaction between two macroscopic bodies. The calculations will use as a model the van der Waals interaction potential integrated for several geometries of interest, as shown in Fig. 13.1 and summarized in Table 13.1.

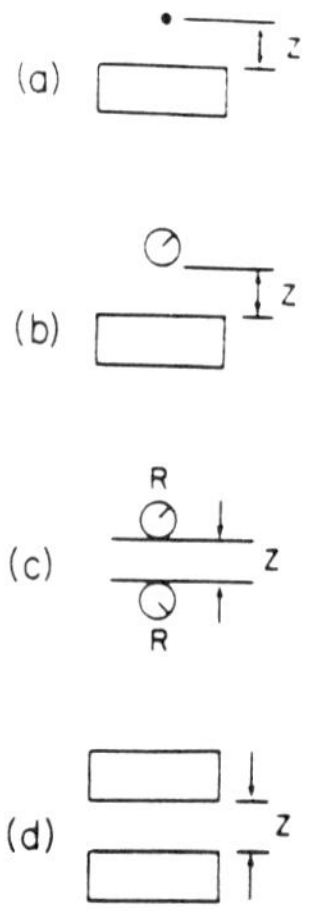

(a) z

(b) z

(c) R z R

(d) z

Fig. 13.1 Geometry of (a) molecule plane, (b) sphere plane, (c) sphere sphere, and (d) plane plane used to model the van der Waals interaction.

Table 13.1. van der Waals interaction energies between macroscopic bodies

	Molecule	Sphere	Surface	Cylinder (↑)
Molecule	$-\dfrac{C}{z^6}$		$-\dfrac{\pi C\rho}{6z^3}$	
Sphere		$-\dfrac{AR'}{6z}$	$-\dfrac{AR}{6z}$	
Surface			$-\dfrac{A\ell^2}{12\pi z^2}$	
Cylinder (↑)				$-\dfrac{A\ell R'^{1/2}}{12\sqrt{2}z^{3/2}}$
Cylinder (→)				$-\dfrac{A(R_1R_2)^{1/2}}{6z}$

$R' = R_1R_2(R_1 + R_2)$
$A = \pi^2 C\rho_1\rho_2 \simeq 10^{-19}$ J = Hamaker constant
ℓ = length

13.3.1. Hamaker Constant

A very useful parameter that appears in many expressions describing the interaction between macroscopic bodies is the Hamaker constant, A, defined by

$$A = \pi^2 C \rho_1 \rho_2 \ . \tag{13.34}$$

Here $C \simeq 5\times10^{-78}$ J m^6 is the coefficient in the molecule-molecule pair potential, Eq. (13.1), and $\rho_i \simeq 3\times10^{28}$ m^{-3} are the number density of molecules in the two interacting media. It happens that the value of A varies only in the narrow range, 0.4×10^{-19} J $< A < 4\times10^{-19}$ J. The reason that A is almost a constant can be understood by considering the proportionality terms

$$A \propto C\rho^2 \propto \alpha^2 \rho^2 \propto \frac{V^2}{V^2} \ , \tag{13.35}$$

where V is the volume in consideration, and α is the polarizability.

13.3.2. Examples of Macroscopic Interactions

We shall now calculate the interaction between several macroscopic bodies of interest. Consider first a point molecule and a planar surface separated by a distance z, shown in Fig. 13.1(a). Using a molecule-molecule pair potential, and assuming that additivity is permissible, we can integrate the interaction between a molecule and a plane surface with a number density of ρ, obtaining, for $n > 3$,

$$W(z) = -\frac{2\pi\rho C}{(n-2)(n-3)} \frac{1}{z^{n-3}} \ , \tag{13.36}$$

which, for $n = 6$, gives

$$W(z) = -\frac{\pi\rho C}{6} \frac{1}{z^3} \ . \tag{13.37}$$

Next, consider a sphere with radius R and number density ρ_1, separated by a distance z from a planar surface having number density ρ_2, as shown in Fig. 13.1(b). For $z << R$ we get

$$W(z) = -\frac{4\pi^2 \rho_1 \rho_2 CR}{(n-2)(n-3)(n-4)(n-5)} \frac{1}{z^{n-5}}, \tag{13.38}$$

and, using $n = 6$, we obtain for the interaction

$$W(z) = -\frac{\pi^2 \rho_1 \rho_2 CR}{6} \frac{1}{z}. \tag{13.39}$$

In terms of the Hamaker constant, the interaction can be written as

$$W(z) = -A \frac{R}{6} \frac{1}{z}. \tag{13.40}$$

For example, a 1-μm-radius sphere separated by a distance of 10 Å from a surface, with $A = 1\times10^{-19}$ J, is attracted by a force of 1.6×10^{-8} N. Assuming that $z \gg R$, we obtain for the sphere-surface interaction

$$W(z) = -2\pi \frac{\rho_1 \rho_2 C V_2}{(n-2)(n-3)} \frac{1}{z^{n-3}}, \tag{13.41}$$

where the volume $V_2 = 4\pi R_2^3/3$. Using $n = 6$ yields for the interaction

$$W(z) = -\frac{\pi \rho_1 \rho_2 C V_2}{6} \frac{1}{z^3}, \tag{13.42}$$

and in terms of the number of molecules, $N_2 = \rho_2 V_2$, the interaction can be written as

$$W(z) = -\frac{\pi \rho_1 C N_2}{6} \frac{1}{z^3}. \tag{13.43}$$

For example, a 25-Å-radius sphere separated by a distance of 250 Å from a plane, with $A = 10^{-19}$ J, is attracted by a force of 3×10^{-15} N, which is negligible. The interaction between two spheres whose radii R_1 and R_2 are much larger than their separation z, as shown in Fig. 13.1(c), is

$$W(z) = -\frac{\pi^2 C \rho_1 \rho_2}{6} \frac{R_1 R_2}{R_1 + R_2} \frac{1}{z}. \tag{13.44}$$

For example, two 1-μm-radius spheres separated by a distance of 10 Å, with $A = 10^{-19}$ J, are attracted by a force of 10^{-8} N. The last example is that of two planar surfaces of number densities, ρ_1 and ρ_2, separated by a distance z, as shown in Fig. 13.1(d). The interaction per area ℓ^2 is given by

$$W(z) = -\frac{2\pi\ell^2\rho_1\rho_2 C}{(n-2)(n-3)(n-4)}\frac{1}{z^{n-4}}, \tag{13.45}$$

which, for $n = 6$, gives for the interaction

$$W(z) = -\frac{A\ell^2}{12\pi}\frac{1}{z^3}. \tag{13.46}$$

By equating the interaction per unit area of two planar surfaces to the interaction of a sphere with a planar surface, for the same separation z, we obtain an effective sphere area

$$A_{eff} = 2\pi R z, \tag{13.47}$$

for $n = 6$. We can, therefore, replace the sphere with a planar surface with area A_{eff} and obtain the same interaction. The interaction between two cylindrical surfaces is left as an exercise.

13.3.3. Derjaguin Approximation

Consider two identical spheres whose radii are much larger than their separation. The force acting between the two is obtained by differentiating the potential with respect to the distance, giving

$$F_s(z) = -\frac{4\pi^2\rho^2 CR}{(n-2)(n-3)(n-4)}\frac{1}{z^{n-4}}. \tag{13.48}$$

Consequently, the relationship between the force acting on the two spheres, $F_s(z)$, and the interaction energy per unit area, $W_a(z)$, of two plane surfaces also separated by z, is given by

$$F_s(z) = 2\pi r W_a(z). \tag{13.49}$$

If the radii are dissimilar, then from a simple geometric argument we find that

$$F_s(z) \simeq 2\pi\frac{R_1 R_2}{R_1 + R_2}W_a(z), \tag{13.50}$$

which is the Derjaguin approximation. Equation (13.50), which holds for $z << R$, is general and can be applied to attractive as well as repulsive forces. However, because

$$F_a(z) = - \partial W_a / \partial z \tag{13.51}$$

and

$$F_s(z) \propto W_a(z) , \tag{13.52}$$

it is possible to have F_s acting in a direction opposite to F_a, that is, an attractive force for two planes can be a repulsive force for two spheres, or vice versa.

13.3.4. Lifshitz Model

The method we have used to calculate the force and interaction energy seems somehow flawed, because we simply have integrated the contribution from single molecules to obtain the total interaction. The correct way to account for these contributions is to solve the problem self-consistently, as is done extensively in electromagnetic theory using the method of images. Consider, for example, the interaction between a charge and a molecule, given by

$$w(z) = - \frac{q^2 \alpha_2}{2(4\pi\epsilon_3)^2} \frac{1}{z^4} , \tag{13.53}$$

where α_2 is the excess polarizability of molecule 2 embedded in medium 3. Integrating this interaction across a surface with density ρ_2 gives

$$W(z) = - \frac{\pi q^2 \rho_2 \alpha_2}{2(4\pi\epsilon_3)^2} \frac{1}{z} . \tag{13.54}$$

Next, we use the method of images to solve the problem, that is, assume that the free charge generates an image charge at the other side of the plane surface. The interaction is then given by

$$W(z) = - \frac{q^2}{4(4\pi\epsilon_3)} \frac{\epsilon_2 - \epsilon_3}{\epsilon_2 + \epsilon_3} \frac{1}{z} , \tag{13.55}$$

where ϵ_i are the dc dielectric constants of the two media. We can now equate Eq. (13.54) with Eq. (13.55) and get for the excess polarizability

$$\alpha_2 = 2\,\frac{\epsilon_3}{\rho_2}\,\frac{\epsilon_2 - \epsilon_3}{\epsilon_2 + \epsilon_3}\,. \tag{13.56}$$

It can be shown that the Hamaker constant for two bodies, with dielectric constants ϵ_1 and ϵ_2 and refractive indices n_1 and n_2, separated by a medium with dielectric constant ϵ_3 and refractive index n_3, is given by

$$A = \frac{3\hbar\omega_e}{8\sqrt{2}}\,\frac{(n^2{}_1-n^2{}_3)(n^2{}_2-n^2{}_3)}{(n^2{}_1+n^2{}_3)^{1/2}(n^2{}_2+n^2{}_3)^{1/2}[(n^2{}_1+n^2{}_3)^{1/2}+(n^2{}_2+n^2{}_3)^{1/2}]}$$

$$+\ \frac{3}{4}\mathrm{KT}\,\frac{\epsilon_1-\epsilon_3}{\epsilon_1+\epsilon_3}\,\frac{\epsilon_2-\epsilon_3}{\epsilon_2+\epsilon_3}\,. \tag{13.57}$$

Here ω_e is the main electronic absorption in the ultraviolet ($\simeq 2\times 10^{16}$ rad/s), n is the refractive index in the visible, and ϵ is the static dielectric constant. The significance of the Lifshitz result is that the Hamaker constant is positive for identical bodies, and can be either positive or negative, depending on the relative values of the parameters. Note that for $A > 0$ we get $w(r) < 0$, and the force involved in the interaction, therefore, is attractive.

13.3.5. Surface Energy

We have seen that the interaction energy between two bodies can be calculated by summing up their van der Waals pair-wise potentials. The summation was performed for any given atom in one body and the rest of the atoms in the other body. If we include in the summation the interaction of each atom with the rest of the atoms in the same body, then for two planar surfaces, for example, we get for the total interaction energy, W_t,

$$W_t(z) = \frac{A}{12\pi}\left[\frac{1}{z_0^2} - \frac{1}{z^2}\right] \times \text{Area}\,. \tag{13.58}$$

Here z_0 is a typical interatomic distance inside each medium, and z is the distance between the two surfaces separating the two bodies. If the two bodies are in contact, then $W_t = 0$, while for large values of z we get

$$W_t(\infty) = \frac{A}{12\pi z_0^2} \times \text{Area}\,. \tag{13.59}$$

The energy required to separate the two bodies, called the surface energy and denoted by 2γ, is given in terms of the Hamaker constant A and the interatomic distance z_0 by

$$\gamma = \frac{A}{24\pi z_0^2} . \tag{13.60}$$

Considering a closely packed solid, it can be shown that the value of z_0 can be approximated by

$$z_0 \simeq \sigma/2.5 , \tag{13.61}$$

where σ is the distance between atomic centers. For solids and many liquids there is, therefore, a relationship between the surface energy and the Hamaker constant, given by

$$\gamma \simeq 5.92\times10^{17}\ \mathrm{A/Area} . \tag{13.62}$$

Because $A \simeq 10^{-19}$ J, we find that the surface energy for many solids and liquids is approximately

$$\gamma \simeq 20\ \mathrm{mJ/m^2} . \tag{13.63}$$

The concept of surface energy can be applied to similar bodies, and the work required to separate them is called the work of cohesion, while for dissimilar bodies the term adhesion is used. The surface energy of a body is roughly proportional to its latent heat and boiling temperature, from which one can predict the relative values for various materials. Surface energy is also a strong function of any thin film deposited on the surface of a body. For example, exposing a mica surface cleaved in vacuum conditions to air, where a thin film of a contaminant can form on it, will reduce the surface energy by an order of magnitude. The adhesion force between two bodies, 1,2 embedded in medium 3, which can be obtained using Derjagun's approximation,

$$F \simeq 2\pi \frac{R_1 R_2}{R_1 + R_2} W_{123} , \tag{13.64}$$

yields for two identical spheres,

$$F = 2\pi R \gamma_l \tag{13.65}$$

and

$$F = 2\pi R \gamma , \tag{13.66}$$

where γ_l and γ denote the surface energy in liquid and in vacuum, respectively. For a sphere and a flat surface we have

$$F = 4\pi R\gamma_l \tag{13.67}$$

and

$$F = 4\pi R\gamma \ . \tag{13.68}$$

The presence of a liquid in the contact area of two bodies, and in particular for two spheres, exerts a force given by

$$F = 4\pi R\gamma_l \cos\theta \ , \tag{13.69}$$

where θ is the contact angle. Because γ of humid air is larger than γ of dry air, the adhesion force will increase upon an increase in the humidity.

13.3.6. Interaction of Particles in Liquids

It is worthwhile mentioning that the interaction of particles in liquids can arise from attractive van der Waals forces, repulsive electrostatic double-layer forces, and attractive or repulsive hydration and steric forces. For a comprehensive review of the topics discussed in sections 13.2 and 13.3, the reader is referred to Israelachvili (1985, 1991).

13.4. Lever-Tip-Sample Contact Interactions

13.4.1. Early Theories

Binnig et al. (1986) estimated the interatomic forces by considering the binding energy of ionic bonds, van der Waals bonds, and those of reconstructed surfaces. The energies of these bonds are on the order of 10 *eV* (1.6×10^{-18} J), 10 meV, and 1 meV, respectively. For distances on the order of 0.16 Å, they calculate respective forces of 10^{-7} N, 10^{-11} N, and 10^{-12} N, respectively. They considered mechanical "springs," composed of levers with their associated tips, that have a mass of 10^{-10} kg and a resonance frequency of 2 kHz, corresponding to a spring constant k = 0.016 N/m. Since an STM can measure displacements as small as 10^{-4} Å, the associated forces acting on the lever will be 1.6×10^{-16} *N*. Thus, they established the feasibility of mapping atomic forces using a tunneling tip together with a macroscopic lever-tip system. At this point in time, Soler et al. (1986), Bando et. al. (1987), Coombs and Pethica (1986), Dürig et. al. (1986), and Pethica (1986) considered the fact that STM images reflect not only the electronic structure at the surface of a sample, but also the deformation induced by the tunneling tip. To analyze this effect, Soler et al. (1986) applied the standard elastic theory to the interaction of a rigid force-sensing tip with radius R and the surface of graphite. The deformation of the surface u was calculated from

$$u = C \int_0^\infty f(r)dr \ , \tag{13.70}$$

where $f(r)$ is the force per unit area, $C = 8\times10^{-11}\ \mathrm{m^2/N}$ for graphite, and r is the distance parallel to the surface. They performed a Fourier analysis and summation of the elastic deformations over the wave vectors, taking into account the compressibility shear between the basal planes of graphite. In the process they approximated $f(r)$ by a Morse potential of the form

$$f(z) = 2bV_0\,[\exp\,[-2b\,(z-z_0)] - \exp\,[-b\,(z-z_0)]] \ , \tag{13.71}$$

with $V_0 = 0.28\ \mathrm{J/m^2}$, $b = 0.97\ \text{Å}^{-1}$, and $z_0 = 3.35$ Å. The resulting deformation of the surface of graphite was

$$u(d) = -1.2\,[\exp\,[-2b\,(d-d_0)] - \exp\,[-b\,(d-d_0)]] \ , \tag{13.72}$$

where

$$z(r) = d + r^2/2R \ , \tag{13.73}$$

with $d_0 = 3$ Å and $R = 2$ Å. Plots of the enhanced vertical displacement, $d + u$, as a function of tip-sample distance d, show a large amplification of the observed corrugation in the repulsive mode.

In an attempt to analyze their experimental images of highly oriented pyrolytic boron nitride (HOPBN), Albrecht and Quate (1987) used a model of a repulsive force assuming that the tip had a single oxygen atom interacting with a single boron or nitrogen on the surface of the sample. They used the Gordon-Kim theory in which the interaction between two neon atoms has been calculated, together with a scaling method devised by Harrison (1981). The overlap potential W_{12} of two atoms, using this model, is given by

$$W_{12}(r) = \eta \hbar^2 r \, \frac{k_{12}^{\;3}}{2m_e} \exp(-5k_{12}r/3) \;, \tag{13.74}$$

where η is an empirical constant and $k_{12} = (1/2)(k_1 + k_2)$. Here

$$k_i = (2m_e W_{p,i})^{1/2}/\hbar \;, \tag{13.75}$$

and $W_{p,i}$ is the energy of atom i in its p-state. To scale the potential of the two neon atoms to that of the oxygen-boron (OB) and oxygen-nitrogen (ON) cases, they used the fact that the charge distribution between the boron and the nitrogen on the surface of the sample is in the ratio 0.7 to 1.3. From this they expressed the OB and ON potentials as

$$W_{12}(r) = (k_{12}/k_{Ne})^3 \, W_{Ne}(r) \exp[-5(k_{12}-k_{Ne})\, r/3] \;. \tag{13.76}$$

Approximating the Gordon-Kim potential by

$$W_{Ne}(r) = A \exp(-\kappa r) \;, \tag{13.77}$$

where A = 2800 eV and κ = 4.45 Å^{-1}, yields the force

$$F_{BO}(r) = -0.7 \, \frac{\partial W_{BO}}{\partial r} = 3.7\times10^{-6} \exp(-3.34\mathrm{r}) \tag{13.78}$$

and

$$F_{NO}(r) = -1.3 \, \frac{\partial W_{NO}}{\partial r} = 11\times10^{-6} \exp(-3.68\mathrm{r}) \;. \tag{13.79}$$

Using this theory they find that for a force of 2×10^{-7} N the tip-sample distances over the boron and nitrogen atoms are 0.88 Å and 1.09 Å, respectively. They note, however, that this theory is not applicable if one has to use interactions involving several atoms.

The Lennard-Jones pair potential has been used to model the interaction of a carbon atom at the apex of a tip interacting with a graphite sample (Pethica and Sutton 1988 and Gould et al. 1989a, for example; see also Gould et al. 1990c). The latter authors applied the many-body potential

$$W = W(\mathbf{r}, \mathbf{r}_1, \mathbf{r}_2, \ldots \mathbf{r}_n) \;, \tag{13.80}$$

where $\mathbf{r}$ is the position of the tip atom and $\mathbf{r}_i$ are the positions of the sample atoms interacting with the tip atom. The solution of the n+1 equations was obtained under the equilibrium condition

$$\frac{\partial W}{\partial \mathbf{z}_i} = 0 , \tag{13.81}$$

from which the force perpendicular to the surface, $F_z = - \partial W/\partial z$, was derived. The constant force measurements give z as a function of the lateral position and the predetermined normal force F_z. The use of a pair potential implies that

$$W(\mathbf{r}, \mathbf{r}_1,\mathbf{r}_2,...,\mathbf{r}_n) = \sum_{i=1} V(\mathbf{r} - \mathbf{r}_i) + V_1(\mathbf{r}_1,...,\mathbf{r}_n) , \tag{13.82}$$

where $V_1(\mathbf{r}_1,...,\mathbf{r}_n)$ is the potential of the sample in the absence of the tip. Using a harmonic approximation they write

$$V_1(\mathbf{r}_1,...,\mathbf{r}_n) = \frac{1}{2} \sum_{i,j,\mu,\nu} u_\mu{}^i \, D_{\mu\eta}{}^{i,j} \, u_\nu{}^j , \tag{13.83}$$

where the $u_\mu{}^i$ and $u_\nu{}^j$ are the displacements of the i and j atoms, μ and ν are the cartesian components, and $D_{\mu\eta}{}^{i,j}$ is the matrix of force constants. Although interatomic energies and forces are not necessarily additive, they used the Lennard-Jones potential in the form of

$$V(|\mathbf{r}_{tip} - \mathbf{r}_{atom}|) = V_0 \left[\frac{1}{2} (r_0/r)^{12} - (r_0/r)^6 \right] , \tag{13.84}$$

with $V_0 = 2.8\times10^{-21}$ J and $r_0 = 2.8$ Å chosen for graphite. The surface layer consisted of six to eight atoms two to three layers deep interacting with the tip atom. The relaxation of these atoms in the presence of the tip atom, using a minimum energy constraint, produced images of constant repulsive force which, under the harmonic approximation, was limited to forces smaller than 10^{-8} N. They find that images obtained with and without relaxation effects are close to each other. The experimental corrugation for graphite, which is the difference in height between site A and the hole, $z(A,h) = 0.2$ Å, could be reproduced by their model. The experimentally measured difference in height of sites A and B was $z(A,B) = 0.02$ Å. Their model, however, gave only a tenth of this value. They found that the asymmetry between the results for sites A and B could be rectified by using double atom tips. By choosing such a tip they could reproduce a variety of images which were compatible with experimental results.

13.4.2. Molecular Dynamics Simulations: Semiconductor-Tip and Semiconductor-Sample Interactions

The Stillinger-Weber (SW) potential (Stillinger and Weber 1985) has been recently used as a realistic model describing tip-sample interactions in molecular dynamics simulations. The original computer simulations using this potential with time units of 7.6634×10^{-14} s have reproduced experimental results of the local order in condensed phases of tetrahedral semiconductors, and in particular, those of silicon. The motive for developing this potential derives from the fact that no reasonable pair potential alone can stabilize the diamond structure, and that a nonadditive interaction potential is required. Specifically, the SW potential consists of two components; a pair potential, $w_2 = \epsilon f_2(\mathbf{r}_{ij}/\sigma)$, and a triplet potential, $w_3(\mathbf{r},\mathbf{r}_j,\mathbf{r}_k) = \epsilon f_3(\mathbf{r}_i/\sigma,\mathbf{r}_j/\sigma,\mathbf{r}_k/\sigma)$, where $\epsilon = 3.4723\times10^{-19}$ J/at pair and $\sigma = 2.0591$ Å. The pair potential is a scalar function, while the triplet potential has to reflect the symmetry of the crystal. The potentials are

$$f_2(r) = A\,[B\,r^{-p} - r^{-q}]\,\exp[(r - \mathrm{a})^{-1}]\,, \qquad (13.85)$$

for $r < \mathrm{a}$, and 0 for $r \geq \mathrm{a}$, and

$$f_3(\mathbf{r}_i\,\mathbf{r}_j\,\mathbf{r}_k) = h(r_{ij},r_{ik},\theta_{jik}) + h(r_{ji},r_{jk},\theta_{ijk}) + h(r_{ki},r_{kj},\theta_{ikj})\,. \qquad (13.86)$$

Also, the function h is given by

$$h(r_{ij},r_{ik},\theta_{jik}) = \lambda\exp\left[\frac{\gamma}{r_{ij}-\mathrm{a}} + \frac{\gamma}{r_{ik}-\mathrm{a}}\right]\left[\frac{1}{3} + \cos\theta_{jik}\right]^2, \qquad (13.87)$$

where θ_{jik} is the angle between $\mathbf{r}_j$ and $\mathbf{r}_k$ subtended at vertex i. The parameters that best fit silicon were found to be: $A = 7.049556277$, $B = 0.6022245584$, $p = 4$, $q = 0$, $a = 1.8$, $\lambda = 21$, and $\gamma = 1.2$. Figure 13.2 shows the energy, force, and force derivative of the SW pair potential which is applicable to a semiconductor sample probed by a semiconductor tip. Note, however, that for distances larger than 3.5 Å, the potentials do not reproduce the long-range van der Waals interaction. We can, therefore, conclude that the SW potential is a useful tool for the analysis of tip-sample interactions in the contact regime that covers the range of 1.5 Å to 3.5 Å. In the range of 2 Å to 3.5 Å the energy, force, and force derivative have roughly the values of $\pm50\times10^{-20}$ J, $\pm50\times10^{-10}$ N, and ±150 N/m.

The SW potential has been used to simulate the effect that the shape of the tip has on the images obtained with atomic force microscopy (Abraham et al. 1988e and Abraham and Batra 1989). Here six different silicon tip shapes were examined, ranging from a tip having one atom at its apex to a tip with four atoms at the apex interacting with a silicon substrate. The calculations, which involved nondestructive forces on the order of 10^{-9} N, were first carried out assuming that both tip and sample were rigid, and then repeated by letting the two reach thermal equilibrium. The interaction forces changed on relaxation by more than an order of magnitude, and their sign may even reverse. It seems, therefore, that any realistic calculation has to include relaxation effects that involve nonadditive interactions. As a result, one cannot use the superposition of images obtained by a tip having a single atom at its apex and obtain images of multi-atom tips. True images of the surface were obtained for only single-atom tips, and multi-atom tips produced complicated artifacts. It was also found that the tip may experience a repulsive force when passing across surface atoms and an attractive force when passing across points between these atoms.

An SW potential has been used to model tip-sample interactions assuming that both tip and sample consist of the same material (Landman et al. 1989a,b). Their calculations were carried out for constant tip-sample height and constant tip-sample force modes of operation. In these calculations the sample consisted of 4 layers, 49 or 100 Si atoms per layer, and the tip consisted of 4 atoms in an initial tetrahedral configuration mounted on 2 layers of Si. The parameters used in the potential were the same as those of SW. In the simulations the tip was lowered slowly toward the sample, and at each position the calculations proceeded until thermal equilibrium was reached. It was found that at a distance of z_1 = 2.91 Å, the force was attractive, at z_2 = 2.345 Å, the force was zero, and at distances below, the force was repulsive. To simulate a constant-force mode, one must include the equation of motion of the lever with its force-sensing tip, as used in Chapter 2. The dynamics of tip-sample interactions could be analyzed by observing trajectories of the atoms during the approach and retract directions of tip motion. On approach and in the repulsive regime, atoms at the top layer of the sample drop to interstitial positions which, on retraction, will anneal. On approach and on retraction they find a localized outward displacement manifesting itself by a connective neck. They observed that the interaction with the tip in the contact regime that generated the interstitial defects modified the interaction force, and sometimes even reversed its direction. The constant-force mode of operation was simulated by using a force of 2.15×10^{-8} N, a tip with 102 atoms, and a surface with 6 layers having 100 atoms per layer. Approach and retraction produced results similar to those found for the constant-height mode.

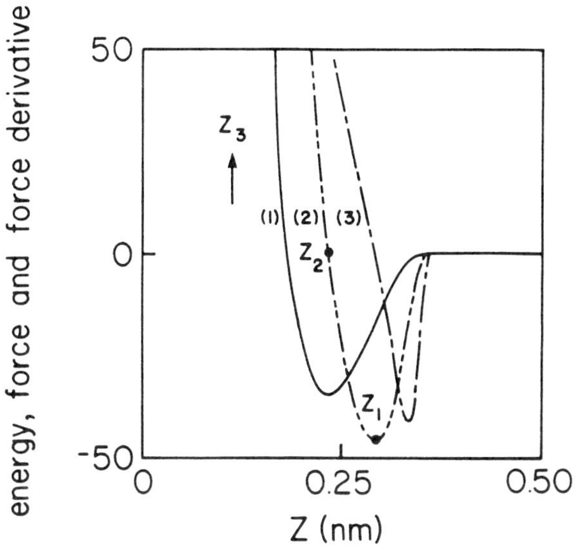

Fig. 13.2. Energy (1), force (2), and force derivative (3) of the SW pair potential that is applicable to a semiconductor sample probed by a semiconductor tip. The distance at which the force attains its maximum attractive value is denoted by z_1, and at z_2 it crosses zero (see Stillinger et al. 1985).

13.4.3. Molecular Dynamics Simulations: Metal-Tip and Semiconductor-Sample Interactions

To simulate the interaction of a tip with a graphite substrate, one can use an intralayer SW potential and an interlayer van der Waals potential (Abraham et al. 1988e, Abraham and Batra 1989). The empirical graphite potential function was used in a molecular dynamics simulation, taking relaxation effects into account. The tip in these simulations either had a single carbon atom at its apex or was attached to a graphite flake, which gave rise to a variety of images. The potential they used had the same functional form as that of SW for the covalent bonding interactions of the intralayer carbon atoms of graphite, with the reduced parameters A = 3.3401393, B = 0.58409783, a = 1.8167332, λ = 17.657014, and γ = 1.2. For the interlayer potential they used a Lennard-Jones potential with σ = 3.234 Å and $\epsilon = 5\times10^{-22}$ J. The sample employed in these simulations consisted of a six-layer slab of graphite with 9×9 primitive cells in the surface. It was found that a monatomic tip senses very similar forces on site A and site B, with the force on the hole site being somewhat lower. It was also found that the single atom tip could image each atom on the surface of graphite and produce a honeycomb structure. However, forces larger than 5×10^{-8} N made the tip pierce the surface of graphite. These results do not reproduce the many different images on graphite and do not explain why high forces are required to reproduce atomic images of graphite. Their conclusion was that it is not a single atom at the tip but rather a graphite flake. To simulate the effect of the flake they used 50 atoms oriented at various angles relative to the graphite sample and could reproduce images obtained in experiments, such as observing only every other atom. Their interpretation of the various images was that they result from misregistry of the two layers. The interaction between the flake and the sample also explained why the forces employed in the experiments were higher than expected for a single atom tip.

13.4.4. Molecular Dynamics Simulations: Metal-Tip and Metal-Sample Interactions

Pair potentials are not accurate enough to describe the energetics of metals in the presence of nonuniformities, leading Daw and Baskes (1983, 1984) to form an alternative approach that they called the embedded-atom model (EAM). The model was applied later by Foiles et al. (1986) to fcc metals such as Cu, Ag, Au, Ni, Pd, Pt, and their alloys. The EAM is a realistic model that is as easy to use as the pair-potential model, and is applicable to metals and their alloys giving the energetics and structure of reconstruction at surfaces, grain boundaries, voids, fracture, segregation of alloys, and adhesion. The idea is to compute the energy of each atom from the energy required to embed it in the local electron density of the rest of the atoms. The electron density, in turn, is the sum of the atomic-electron densities, which is assumed to give a constant background. The total energy of the system consists of a core-core pair-repulsion term ϕ and an embedding potential E_i. The repulsive interaction is given by

$$\phi_{ab}(r) = \frac{Z_a(r) Z_b(r)}{r} , \qquad (13.88)$$

with the effective charge density $Z(r)$:

$$Z(r) = Z_0 [1 + \beta r^{\nu}] \exp(-\alpha r) , \qquad (13.89)$$

where a and b refer to the type of the atoms. The embedding functional F gives the energy required to embed atom i in the background electron density ρ of the atomic system and is independent of the source of background electron density. It requires the knowledge of the host electron density $\rho_{h,i}$ at the position of atom i due to the rest of the atoms. The density is computed from Hartree-Fock wave functions using

$$\rho^a(r) = n_s \rho_s(r) + n_d \rho_d(r) , \tag{13.90}$$

where n_s and n_d are the number of the outer s and d electrons, and ρ_s and ρ_d refer to the charge density of the wave functions. The total electron density at atom i is given as a superposition of the charge densities of atoms j

$$\rho_{h,i} = \sum_{j \neq i} \rho_j{}^a(r_{ij}) . \tag{13.91}$$

The many-body functional F is obtained empirically by fitting the total cohesive energy of the fcc solid,

$$W_{tot} = \sum_i E_i(\rho_{h,i}) + \frac{1}{2} \sum_i \sum_{j \neq i} \phi_{ij}(r_{ij}) , \tag{13.92}$$

to a universal equation of state,

$$W(\mathrm{a}) = -W_{sub}(1 + a')\exp(-a') . \tag{13.93}$$

Here W_{sub} is the sublimation energy and a' is a measure of the deviation from the equilibrium lattice constant given by

$$a' = (a/a_0 - 1)\sqrt{9B\Omega/E_{sub}} . \tag{13.94}$$

Here a and a_0 are the length scale characteristics of the condensed phase and the equilibrium lattice constant, respectively, B is the bulk modulus of the material, and Ω is the equilibrium volume per atom. Parameters for the fcc metals obtained from experiments can be used to construct the functions ϕ and F, from which the energetics of these metals and their alloys can be calculated. The embedded-atom model is useful for the calculation of the interaction of a metal tip and a metal sample, and Landman et al. (1990) have used it to calculate the dynamics of adhesion, nano-indentation, and fracture for a nickel tip and gold surface. In the simulations the gold sample consisted of 3 static and 8 dynamic layers, each consisting of 450 atoms. The Ni tip consisted of 1400 atoms arranged originally as a clipped pyramid, where the small area that faces the sample consisted of 72 atoms. Their work successfully reproduced the dynamics of the tip approach and subsequent retraction from the sample and the permanent changes that take place during this process. At a small tip-sample distance, the mechanical instability resulted in a jump to contact from 4.2 Å to 2.1 Å for a tip mounted on a lever with k = 5000 N/m, manifested by an elastic deformation of the sample. Forcing the tip closer to the sample gives rise to an indentation, or permanent plastic deformation of the surface of the sample. Upon separation, the bonding between the atoms at the surface of the tip and the sample gives rise to an adhesion that creates a connective neck and a wetting of the nickel tip by gold atoms, as shown in Fig. 13.3. The simulation also provided the important information regarding the hydrostatic pressure between the tip and the sample, which can be as large as 10^{10} N/m^2, and the resulting dynamic change of the tip structure. For other molecular dynamics simulations, see Paik et. al. (1990), Pethica and Sutton (1990), Zhong and Tománek (1990), and Zhong et. al. (1990).

13.4.5. Classical Models of Indentation, Adhesion, and Tribology

The effective spring constant k' of a system can be generalized to include not only the lever and its force-sensing tip, but also the deformation of the sample, and can be written as

$$k' = k - F'_a + F'_d \ . \tag{13.95}$$

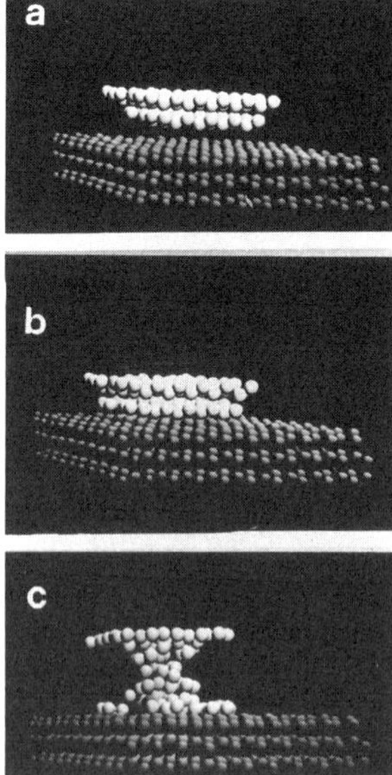

Fig. 13.3 Creation of a connective neck and the wetting of a nickel tip by gold atoms upon tip-sample detachment process. (Courtesy: Uzi Landman, Georgia Institute of Technology.)

Here k is the spring constant of the free lever, F'_a is the derivative of the tip-sample interaction force, and F'_d is the derivative of the force associated with the deformation of the sample. The force derivative associated with the deformation of a sample by the tip can be derived for any tip shape using classical models. Consider, for example, a flat sample deformed by a non-adhesive flat-ended cylindrical tip with radius R acting on it with a force F_d. The deformation z_d of the sample under the action of the tip is (Timoshenko and Goodier 1970)

$$z_d = \frac{1}{2} \frac{1-\nu^2}{E} \frac{1}{R} F_d \,, \tag{13.96}$$

with the force and force derivatives given, respectively, by

$$F_d = 2 \frac{E}{1-\nu^2} R \, z_d \tag{13.97}$$

and

$$F'_d = 2 \frac{E}{1-\nu^2} R \,, \tag{13.98}$$

where ν is Poisson's ratio. Another example is that of a paraboloid-shaped tip that will deform a surface (Cohen et. al. 1990a) according to

$$z_d = \left[\frac{3}{8} \frac{1-\nu^2}{E} \right]^{2/3} \frac{1}{R^{1/3}} F_d^{\,2/3} \,. \tag{13.99}$$

Here the force and force derivative are given, respectively, by

$$F_d = \frac{8}{3} \frac{E}{1-\nu^2} R^{1/2} z^{3/2} \tag{13.100}$$

and

$$F'_d = 4 \frac{E}{1-\nu^2} R^{1/2} z^{1/2} \,. \tag{13.101}$$

An important situation occurs when the effective spring constant vanishes, which takes place when the tip approaches the sample. Using the van der Waals attractive interaction, we find for a non-adhesive flat-ended cylindrical tip and for a paraboloid-shaped tip that this condition is fulfilled, respectively, when

$$k - F'_a + 4 \frac{E}{1 - \nu^2} R^{1/2} z^{1/2} = 0 \tag{13.102}$$

and

$$k - F'_a + 2 \frac{E}{1 - \nu^2} R = 0 \, . \tag{13.103}$$

For a sufficiently soft lever, the elasticity of the sample can be ignored, and the attractive van der Waals interaction will force the lever to jump toward the sample when $k' = 0$, which occurs at a distance $z_{j\,1}$. Neglecting repulsive force, we get

$$z_{j\,1} = \left[\frac{AR}{3k} \right]^{1/3} , \tag{13.104}$$

for a spherical tip, and the attractive force is

$$F_{j\,1} = \frac{1}{2} \left[\frac{ARk^2}{3} \right]^{1/3} . \tag{13.105}$$

The tip-sample interaction force must have a repulsive term to prevent the tip from penetrating the sample. As we shall see later, the combination of attractive and repulsive interactions will generate a second solution to $k' = 0$, which takes place at a tip-sample distance of $z_{j\,2}$, larger than $z_{j\,1}$. In other words, pulling the lever away from the surface will generally result in a second jump, which occurs at a tip-sample distance where the attractive force obtains its maximum value. The simple adhesion theory (Derjaguin et al. 1975) identifies the maximum attractive force with the adhesion force, yielding

$$kz_{j\,2} = 4\pi R \sqrt{\gamma_1 \gamma_2} \, . \tag{13.106}$$

We can also assess the pressure exerted by the tip on the sample using the adhesion theory (Derjaguin et al. 1975) by dividing the measured force by the area of contact. For a sphere and a plane surface, in the absence of external forces, the radius of the contact area is

$$R_c = \left[4\pi \frac{R^2}{k} \sqrt{\gamma_1 \gamma_2} \right]^{1/3} , \tag{13.107}$$

and for a flat surface pressed by a nonadhesive flat-ended cylindrical tip, the radius is

$$R_c = \frac{1}{2}\frac{1-\nu^2}{E}\frac{F_z}{z_s}, \tag{13.108}$$

where z_s is the deformation of the surface. We can define, for the second case, a spring constant k_s associated with the deformation of the surface by

$$k_s = \frac{2ER_c}{1-\nu^2}. \tag{13.109}$$

Recently, Chen and Hamers (1991) investigated the apparent barrier height, as measured by an STM theoretically and experimentally, for a tungsten tip and a silicon sample under UHV conditions. They find that the conductance G between the tip and sample in the STM is related to their adhesion energy U by $U \propto -\sqrt{G}$. If we let z_e be the tip-sample separation when the force is zero, the interaction force could then be written in terms of a Morse potential,

$$F = -2\kappa U\,[\exp(-\kappa(z_s - z_e)) - \exp(-2\kappa(z_s - z_e))]\,, \tag{13.110}$$

with $U = 5$ eV, $E = 3.4\times 10^{11}$ N/m², $\kappa = \sqrt{2m_e\phi/\hbar^2} \simeq 1$ Å, and ϕ is the work function. They were able to verify Eq. (13.110) experimentally and explain the lowering of the barrier quantitatively by taking into account the deformation of the surface. Equation (13.110) can, therefore, be used to model the atomic force microscope operating in the contact mode.

A survey of these and related topics, including other references, can be found in Mate et al. (1987, 1988, 1989a, b, 1990), Ribarsky and Landman (1988), Erlandsson et. al. (1988b), Tang et. al. (1988), Burnham and Colton (1989), Dürig and Züger (1989), Landman et al. (1989a, b and 1990), Taubenblatt (1989), McClelland and Cohen (1990), Cohen et al. (1990a, b), Heinzelmann et al. (1990), Neubauer et al. (1990), Zhong and Tománek (1990), Zhong et. al. (1990), Andoh et. al. (1990), Blackman et. al. (1990), Burnham et. al. (1990), Terris et. al. (1990a,b), Gould et. al. (1990c), Sutton and Pethica (1990), Chen (1991), Ciraci et al. (1992), Hartmann (1992), Ciraci (1992), Chen and Hamers (1991), Bozzolo and Ferrante (1992), Bouju et al. (1992), P. Gleyzes et al. (1991), and Goodman and Garcia (1991).

13.5. Lever-Tip-Sample Noncontact Interactions

Atomic force microscopy provides a means for imaging atomic forces acting between the force-sensing tip and the surface of a sample. We have seen that molecular dynamics simulation can provide a detailed description of the processes taking place when the tip and the sample are in contact, both during approach and retraction of the tip. The noncontact mode is characterized by weak tip-sample interactions which do not modify the interacting bodies, and by a tip-sample distance that is large enough so that atomic details are smeared out. Under these conditions we can use simple analytic expressions for the tip-sample interactions. Specifically, we will examine two topics in this section. The first describes the dynamics of the system consisting of the lever and its force-sensing tip and a sample, which exhibits a bistable behavior with a hysteresis loop. The second concerns a comparison of atomic, electric, and magnetic interactions that can exist simultaneously in experimental situations.

13.5.1. Dynamics of Lever-Tip-Sample Interaction

To describe the dynamics of the tip approach, we have to choose a suitable tip-sample potential that will be incorporated into the total lever-tip-sample interaction. For simplicity we choose the Lennard-Jones potential because of its representation by power-law terms. The Lennard-Jones potential strictly applies only to the interaction between two molecules. We can, however, approximate actual macroscopic interactions by integrating the interaction across microscopic bodies, as discussed in section 13.3.2. For our applications, we choose a sphere representing the force-sensing tip, and a planar surface representing the sample. We take the common form of the potential, neglecting retardation effects,

$$w(r) = 4\epsilon\left[\frac{\sigma^{12}}{r^{12}} - \frac{\sigma^6}{r^6}\right], \tag{13.111}$$

and write the attractive and repulsive components

$$w_6(r) = -\frac{4\epsilon\sigma^6}{r^6} = -\frac{C_6}{r^6} \tag{13.112}$$

and

$$w_{12}(r) = \frac{4\epsilon\sigma^{12}}{r^{12}} = \frac{C_{12}}{r^{12}}. \tag{13.113}$$

Using $C_6 = 4\epsilon\sigma^6$ and $C_{12} = 4\epsilon\sigma^{12}$ in Eq. 13.38, we get for the attractive and repulsive components

$$W_6(z) = -\ 4\pi^2\epsilon\rho_1\rho_2\sigma^5 R\ \frac{\sigma}{6z} \tag{13.114}$$

and

$$W_{12}(z) = 4\pi^2\epsilon\rho_1\rho_2\sigma^5 R\ \frac{1}{1260}\ \frac{\sigma^7}{z^7}\ , \tag{13.115}$$

respectively. The total potential is, therefore,

$$W(z) = -\frac{2}{3}\ \pi^2\epsilon\rho_1\rho_2\sigma^5 R \left[\frac{\sigma}{z} - \frac{1}{210}\ \frac{\sigma^7}{z^7}\right], \tag{13.116}$$

with the force and force derivative given, respectively, by

$$F(z) = -\frac{2}{3}\ \pi^2\epsilon\rho_1\rho_2\sigma^4 R \left[\frac{\sigma^2}{z^2} - \frac{1}{30}\ \frac{\sigma^8}{z^8}\right] \tag{13.117}$$

and

$$F_1(z) = \frac{4}{3}\ \pi^2\epsilon\rho_1\rho_2\sigma^3 R \left[\frac{\sigma^3}{z^3} - \frac{2}{15}\ \frac{\sigma^9}{z^9}\right]. \tag{13.118}$$

We shall use these integrated forms, shown in Fig. 13.4, to analyze the dynamics of tip-sample interactions in atomic force microscopy. The system that we consider consists of a sample, a lever holder, and a lever with its force-sensing tip attached to it. As before, we note the position of the lever in the absence of forces by u, the position of the tip during the interaction by z, and the sample position by $g = 0$. We consider equilibrium conditions where the tip-sample force is equal to the restoring force of the lever at all times, namely,

$$k(z - u) = F_0 \left[\frac{\sigma^2}{z^2} - \frac{\sigma^8}{30z^8}\right], \tag{13.119}$$

where

$$F_0 = \frac{2}{3}\ \pi^2\epsilon\rho_1\rho_2\sigma^4 R\ . \tag{13.120}$$

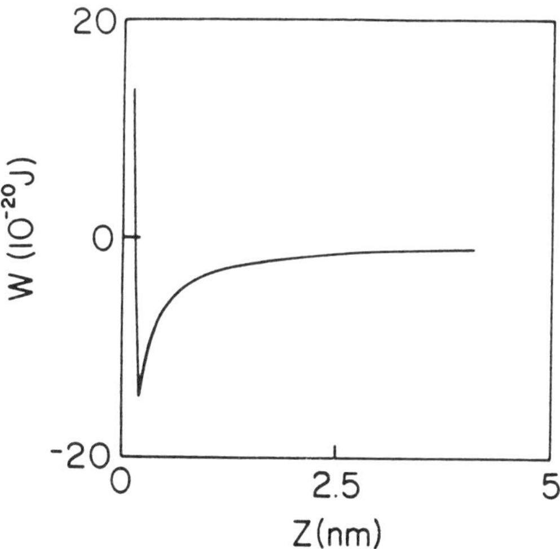

Fig. 13.4 Energy as a function of distance used as an example for a van der Waals pair potential.

In approaching the sample, we start with a large distance u and decrease it gradually, and wish to find the tip-sample distance z as a function of u. However, it is more convenient mathematically to calculate u as a function of z to avoid a cumbersome solution, and to that end we write u as

$$u = z - \frac{F_0}{k}\left[\frac{\sigma^2}{z^2} - \frac{\sigma^8}{30z^8}\right], \tag{13.121}$$

which has a simple functional form. Of interest to our discussion is the total energy of the system, W_t, consisting of that of the tip-sample interaction and that of the potential energy of the bent lever,

$$W_{total} = \frac{1}{2}\,k(z - u)^2 + F_0\sigma\left[\frac{\sigma^7}{210z^7} - \frac{\sigma}{z}\right]. \tag{13.122}$$

A typical choice of parameters that generates a hysteresis loop is given by σ = 3.41 Å, ϵ = 3.79×10^{-22} J, R = 20 Å, and $\rho_1 = \rho_2 = (2.5)^{-3}$ Å^{-3} (Meyer et al. 1988, 1989a). Figure 13.5 shows the total energy of the system for k = 0.1 N/m and u = 10, 15, 20, 25, and 30 Å as a function of z. For each value of u we scan the possible values of z and identify the lowest points on the curve describing the total energy of the system. At equilibrium, the tip will obviously choose the minimum energy and move to its respective value of z. For u = 30 Å, for example, we find a single energy minimum at z somewhat smaller than 30 Å. For u = 10 Å, there is again a single minimum of the energy, but this time at about 2 Å. For $10 < u < 20$ Å, however, we notice that there are two extrema, where one is always slightly lower than the other, and the tip will jump from one to the other when u is slightly shifted. The hysteresis loop is shown in Fig. 13.6, created by plotting the bending of the lever, $z - u$, as a function of u. The arrows relate to the direction of the evolution of the system on changing u. We start the tip approach at point (1) and decrease u. On passing from point (2) to point (3), the tip encounters two extrema in the total energy and will stay at the lowest value. At point (3) the tip finds a deeper value of the energy and will jump to point (4). A further decrease in the value of u will bring the tip closer to the sample, at point (5), and on retraction the tip will follow the trajectory leading to point (4). From here, and up to point (6), the tip again senses two extrema, and at point (6) will jump to the deepest value of the energy which is at point (2). Further increase in u will bring the tip back to point (1) where the whole process started. The region spanned by the values of u, described by points (2) and (3), or by points (4) and (6), has two extrema. For the tip to go from one to the other, it has to climb a small energy barrier separating the two extrema. It will, therefore, stay in one of these, which is not necessarily the lowest, until the potential barrier is lowered so it can "roll" out. Figure 13.6 plays a major role in the understanding of the operation of any atomic force microscope, in spite of the fact that it uses crude approximations. For example, strong deformation of the surface of the sample, and the presence of contaminants on top of it, play a major role in the tip-sample interaction, which we have so far ignored. Nevertheless, experimental curves do exhibit a behavior similar to that shown in Fig. 13.6 with values of u and z close to those described by points (1) to (6). The role of the spring constant in the tip-approach process is shown in Fig. 13.7 for k = 0.1, 0.2, 0.3, 0.4, and 0.5 N/m. We see that the area of the loop describing the bistable region, where there are two extrema, decreases as k decreases. For k values that are large enough, there will always be only a single minimum of the energy and no "jump" will take place in the approach process. The hysteresis loop in this example has a typical shape, which in practice can vary considerably depending on the choice of parameters. In particular, surface energies of the tip and sample will play a major role in the interaction, and the existence of even submonolayers of contaminants on either tip or sample can modify the loop considerably. For a discussion of tip-sample interaction see Pethica and Sutton (1988).

Experiments showing hysteresis loops have been reported, for example, by Anders and Heiden (1988) with k = 1000 N/m, Burnham and Colton (1989) and Burnham et al. (1989) with k = 50 N/m, Cohen

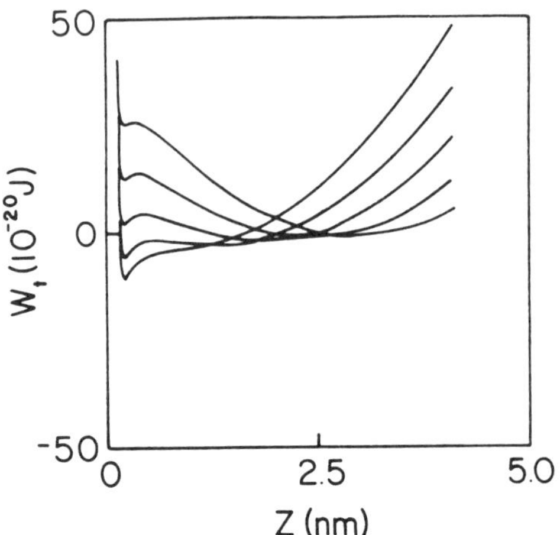

Fig. 13.5 Total energy of a tip-sample system.

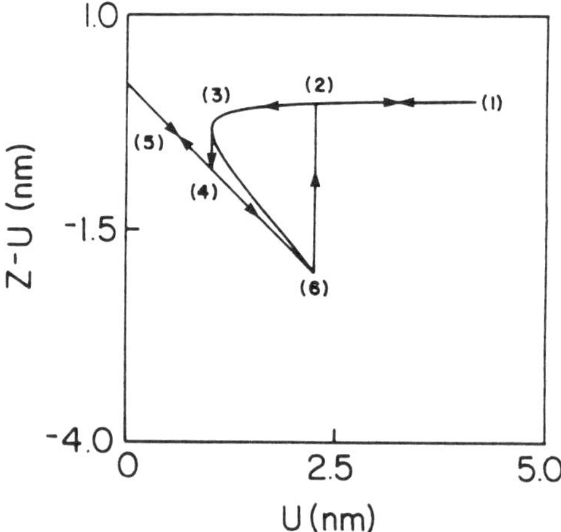

Fig. 13.6 Probe-lever distance, z - u, as a function of probe-sample distance u showing the hysteresis loop.

et al. (1990a,b) with k = 60 N/m and 115 N/m, Erlandsson et al. (1988a,b) with k = 30 N/m, Mate et al. (1987) with k = 150 N/m and 2500 N/m, Mate et al. (1988) with k = 160 N/m, Meyer et al. (1988) with k = 0.1 N/m and 3.8 N/m, and Weisenhorn et al. (1989) with k = 3 N/m.

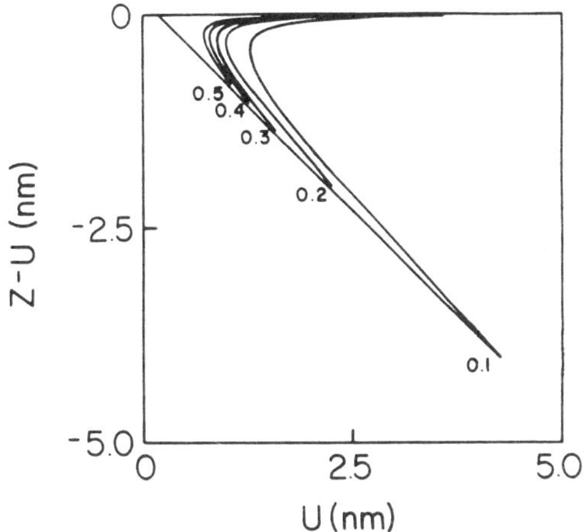

Fig. 13.7 The hysteresis loop for spring constants k = 0.1, 0.2, 0.3, 0.4, and 0.5 N/m.

13.5.2. Comparison of Atomic, Electric, and Magnetic Interactions

It is instructive to compare the interaction energy, force, and force derivative associated with atomic, electrostatic, and magnetostatic noncontact interactions, because in many cases, two, or even all three, participate at the same time. The simplest model for these three interactions will involve a spherical tip with radius R at a distance r from a sample. For the atomic and electric interactions, the sample will be modeled as an infinite plane surface bounding a half space. For this geometry, the atomic interaction using the van der Waals interaction is

$$W_a(z) = -\frac{A}{6}\frac{1}{z'}, \tag{13.123}$$

with the scaled distance z' given by

$$z' = \frac{r}{R}. \tag{13.124}$$

Retardation effects (Israelachvili 1972) will kick in at a distance of approximately 100 Å and will have a transition distance to a larger power coefficient that extends to 1000 Å. For simplicity we neglect this effect, because we seek only the order of magnitude of the interactions. For detailed calculations considering the geometry and material of the tip and sample, see Hartmann (1990a). The electrostatic interaction for the sphere-plane geometry is

$$W_e(z) = 2\pi\epsilon_0 RV^2 \sum_{n=2}^{\infty} \frac{\sinh(\alpha)}{\sinh(n\alpha)} , \qquad (13.125)$$

where

$$\alpha = \log[1 + z' + \sqrt{z'^2 + 2\, z'}] , \qquad (13.126)$$

neglecting the self energy of the sphere. For the magnetostatic interaction we cannot use for the sample an infinite magnetic thin plane film bounding a half space that is uniformly magnetized, because the field outside the surface will be constant and will not exert a force on a magnetic dipole. We therefore replace the surface with a second identical sphere that is uniformly magnetized. Each sphere, representing the tip magnetic domain or the domain on the sample, can now be replaced by a single magnetic dipole placed at the center of each sphere, with a dipole moment given by $m = 4\pi R^3 M/3$, where M is the magnetization of each sphere. The magnetostatic interaction of two dipoles is

$$W_m(r) = \frac{\mu_0}{4\pi}\left[\frac{\mathbf{m}_1\cdot\mathbf{m}_2}{r^3} - \frac{3\,(\mathbf{m}_1\cdot\mathbf{r})(\mathbf{m}_2\cdot\mathbf{r})}{r^5}\right] , \qquad (13.127)$$

and, assuming that their equal moments are aligned perpendicular to the vector connecting their centers, we get

$$W_m = \frac{4\pi}{9}\,\mu_0\,\frac{R^3}{(2 + z')^3}\,M^2 . \qquad (13.128)$$

Note that in each of the three cases the distance z could be written as a scaled parameter. Figures 13.8 and 13.9 show the force and force derivative as a function of distance for $A = 3{\times}10^{-19}$ J, $M = 100$ kA/m, $V = 1$ V, and $R = 1000$ Å. For the atomic force in the range of $50 < z < 200$ Å, we find that $2 < F < 0.12{\times}10^{-10}$ N and $0.07 < F_1 < 0.001$ N/m, scaling as z^{-2} and z^{-3}, respectively. Martin and Wickramasinghe (1987) find for this force, in the range $30 < z < 180$ Å, that $15 < F < 0.1\ 10^{-10}$ N and $2 < F_1 < 0.01$ N/m. The difference between the theoretical and experimental values may be due to the scaling of the absolute tip-sample distance, retardation effects, or different tip shape and diameter. For the electrostatic force in the range of $50 < z < 200$ Å, we find that $2.5 < F < 1.1{\times}10^{-10}$ N and $0.0063 < F_1 < 0.001$ N/m. Martin et al. (1987) approximated the sphere-plane capacitance by that of a plane-parallel capacitor. Since $z < R$, we cannot make an exact comparison of their results with our theory. For z = 100 Å, their force scales to F = 2×10^{-10} N, while theory predicts 5×10^{-10} N. Schönenberger and Alvarado (1990b) measured a force in this range that scales to approximately 10^{-11} N for a tip radius they estimate to be 400 Å, which is within the ballpark of the theory. The difference between experiment and theory may be attributed to the same factors as discussed before. For the magnetostatic force in the range of $50 < z < 200$, we find that $F \simeq 0.3\times10^{-1}$ N and $F_1 \simeq 0.0005$ N/m. The flat curves in this region are due to saturation effects which were measured by Schönenberger and Alvarado (1990a). Martin et al. (1987), however, measured a magnetic force and its derivative as the tip-sample distance decreases from 1400 to 200 Å. No comparison can be made here because the structure of the magnetic domains in their experiments differed from our model, yet the magnitude of their forces is close to that of the theory. It is interesting to note that the van der Waals and magnetostatic forces, using our models and parameters, are equal at a tip-sample distance of 150 Å, which may limit the maximum resolution of magnetic force microscopy; see also Saenz et al. (1987, 1988), and Wadas et al. (1990b). For limits of macroscopic resolution of STM and AFM tips, see Stedman (1988), Anders and Heiden (1990), Hartman (1990b), Reis et al. (1990), Suzuki et al. (1991), Weihs et al. (1991), Giessibl (1992), Moiseev et al. (1990), and Girard (1991). A discussion of electrostatic and capillary forces is given by Hao et al. (1990). Images using atomic forces in the noncontact mode have been obtained, for example, by Martin et al. (1987).

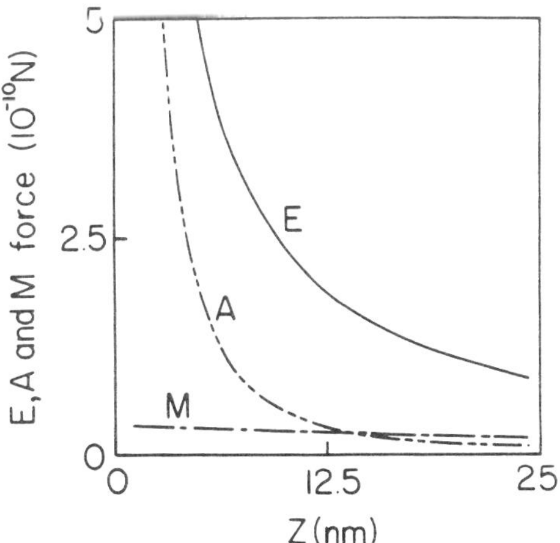

Fig. 13.8 Force as a function of distance for atomic (A), electrostatic (E), and magnetic (M) tip-sample interactions.

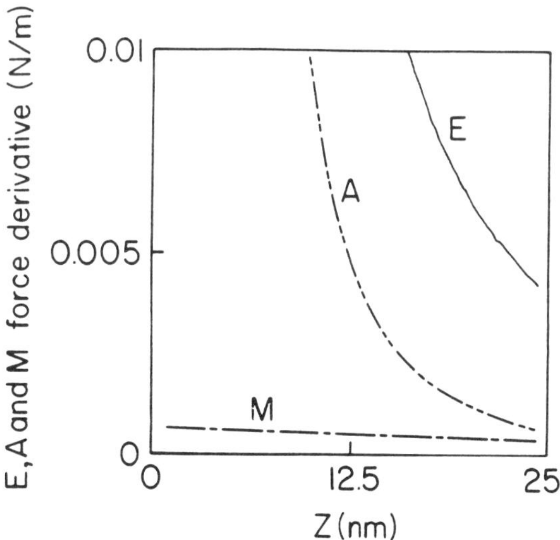

Fig. 13.9 Force derivative as a function of distance for atomic (A), electrostatic (E), and magnetic (M) tip-sample interactions.

13.6. Experimental Results For The Contact Mode

In this section we summarize results of experiments with atomic force microscopy that have been conducted in the contact mode. Figures 13.10 to 13.25 show images produced by atomic force microscopy, demonstrating its power and versatility.

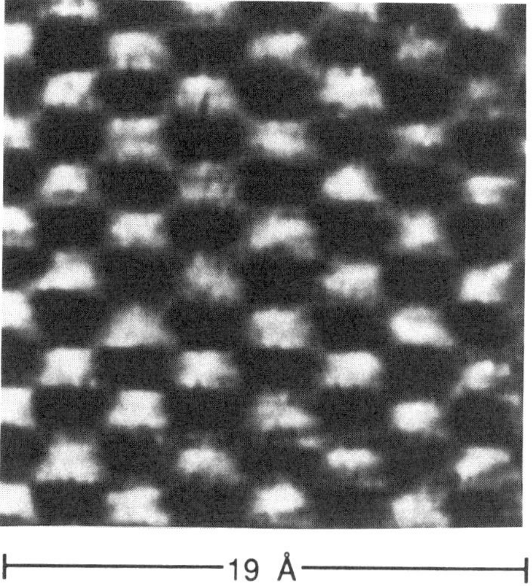

Fig. 13.10 AFM image of graphite. (Courtesy: T. R. Albrecht and C. F. Quate, Stanford University.)

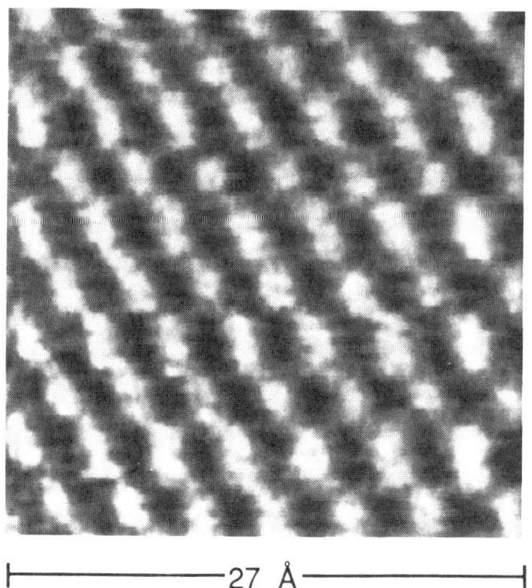

Fig. 13.11 AFM image of MoS_2. (Courtesy: T. R. Albrecht and C. F. Quate, Stanford University.)

Fig. 13.12 AFM image of boron nitride. (Courtesy: T. R. Albrecht and C. F. Quate, Stanford University.)

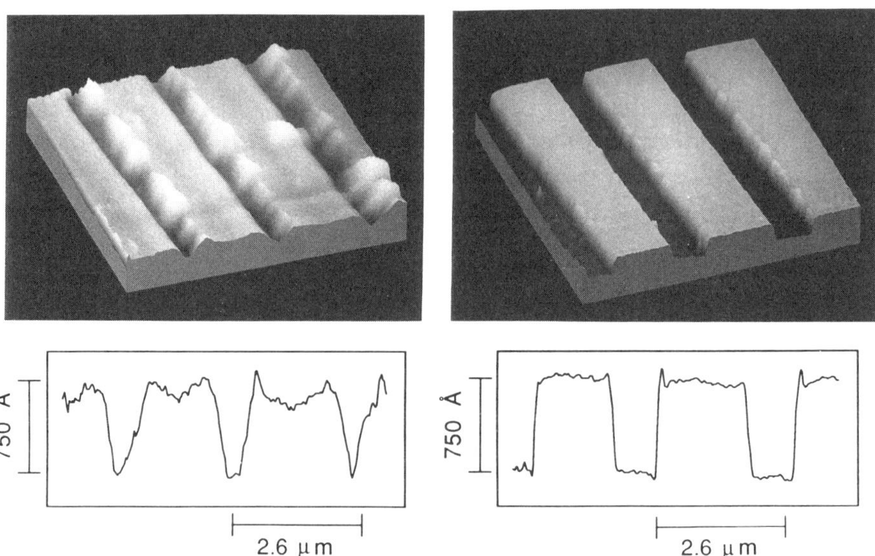

Fig. 13.13 AFM image of a glass diffraction grating with (left-hand side) a flat lever and (right-hand side) a lever with a sharp tip. (Courtesy: S. Akamine, R. C. Barrett, and C. F. Quate, Stanford University.)

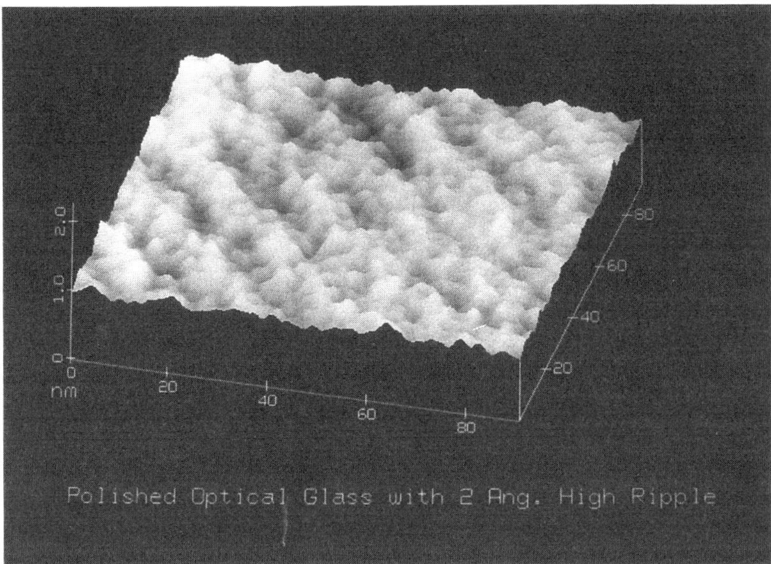

Fig. 13.14 AFM image of an optical glass. (Courtesy: V. Elings, Digital Instruments.)

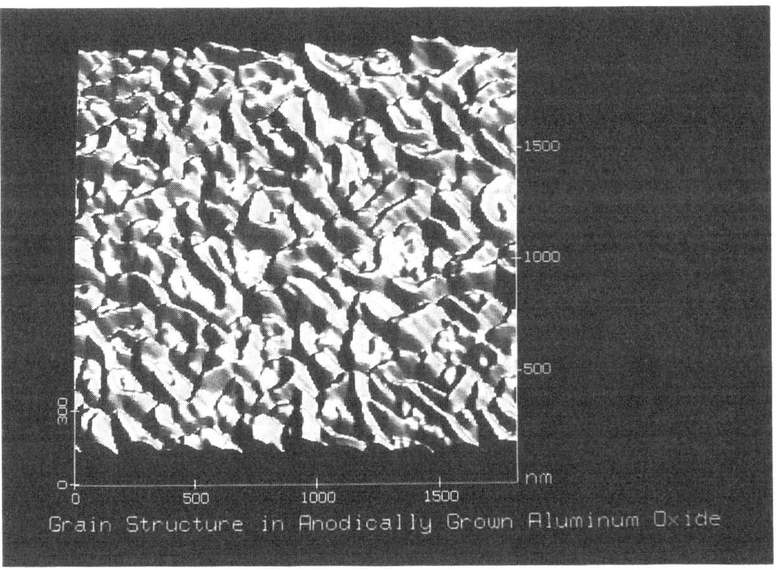

Fig. 13.15 AFM image of aluminum oxide. (Courtesy: V. Elings, Digital Instruments.)

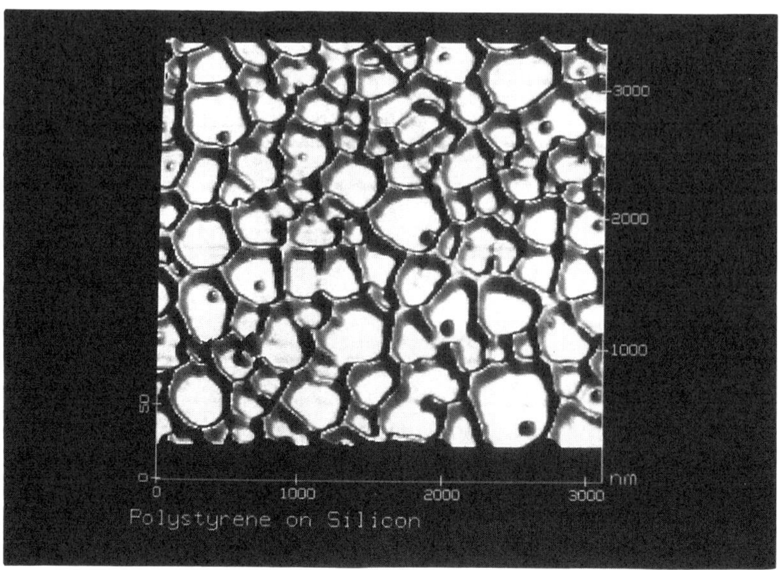

Fig. 13.16 AFM image of polystyrene on silicon. (Courtesy: V. Elings, Digital Instruments.)

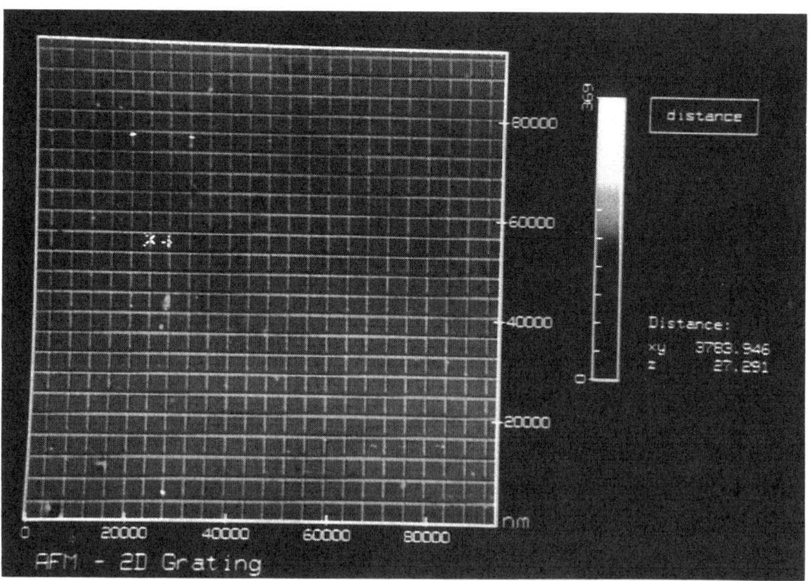

Fig. 13.17 AFM image of a grating. (Courtesy: V. Elings, Digital Instruments.)

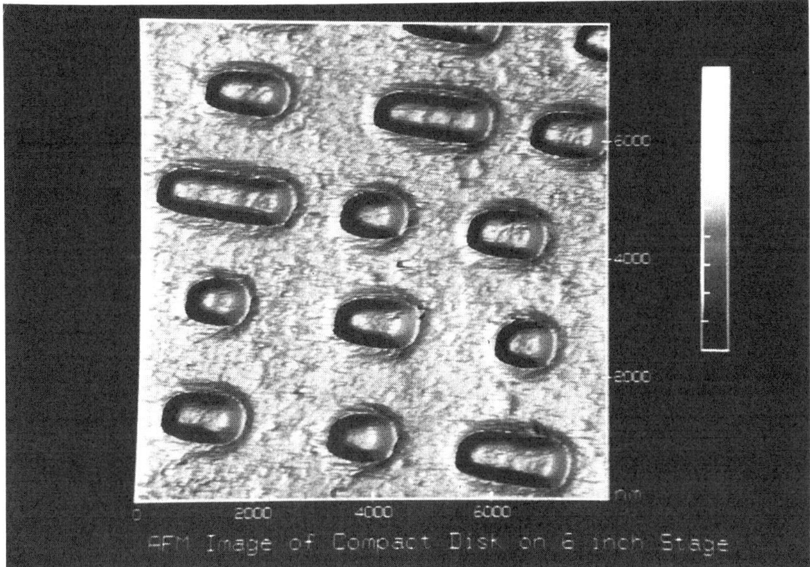

Fig. 13.18 AFM image of a compact disk stamper. (Courtesy: V. Elings, Digital Instruments.)

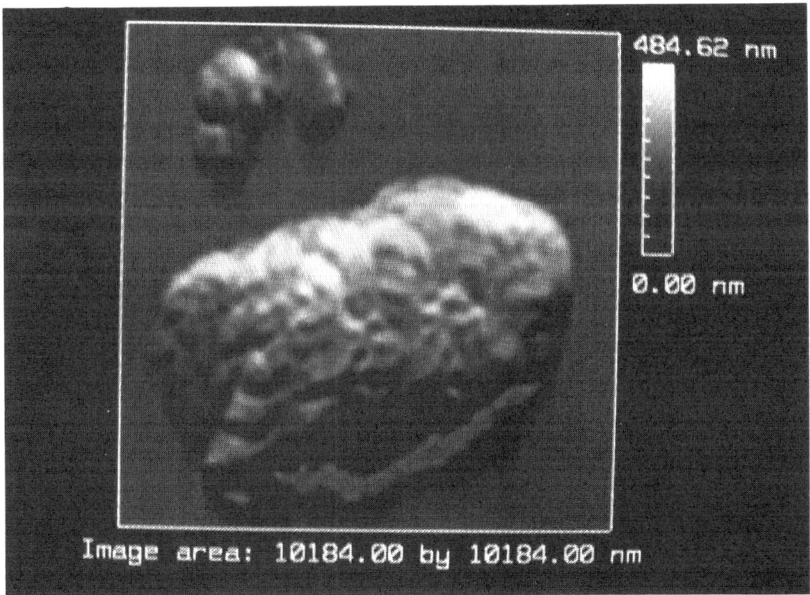

Fig. 13.19 AFM image of a human lymphocite attached to Anti-IgG on a glass substrate. (Courtesy: S. A. Gould, B. Drake, C. B. Prater, A. L. Weisenhorn, S. Manne, H. G. Hansma, and P. K. Hansma, University of California; J. Massie, M. Longmire, and V. Elings, Digital Instruments; B. Dixon Northern, B. Mukergee, C. M. Peterson, Sansum Medical Research Foundation; W. Stoeckenius, Cardiovascular Research Institute; and T. R. Albrecht and C. F. Quate, Stanford University.)

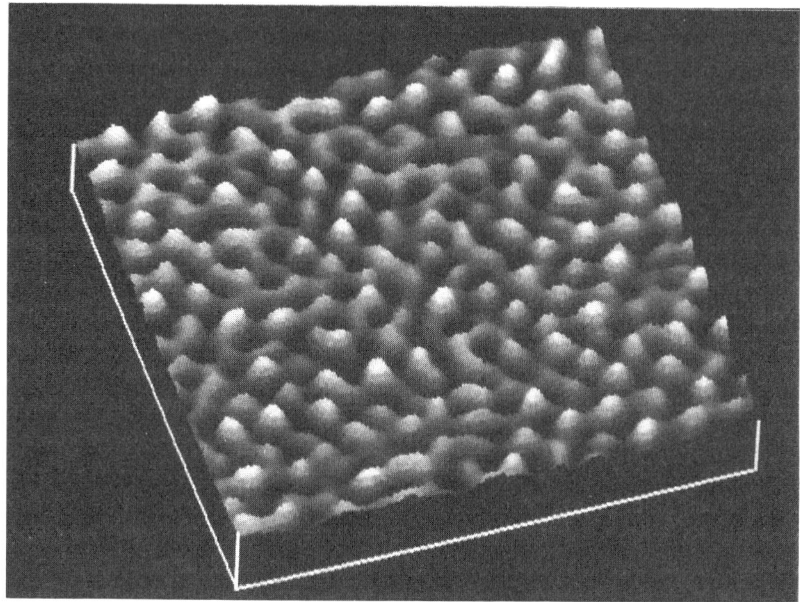

Fig. 13.20 AFM image of gold atoms in air. (Courtesy: S. Manne, H. J. Butt, S. A. C. Gould, and P. K. Hansma, University of California.)

Fig. 13.21 AFM image of gold atoms in water. (Courtesy: S. Manne, H. J. Butt, S. A. C. Gould and P. K. Hansma, University of California.)

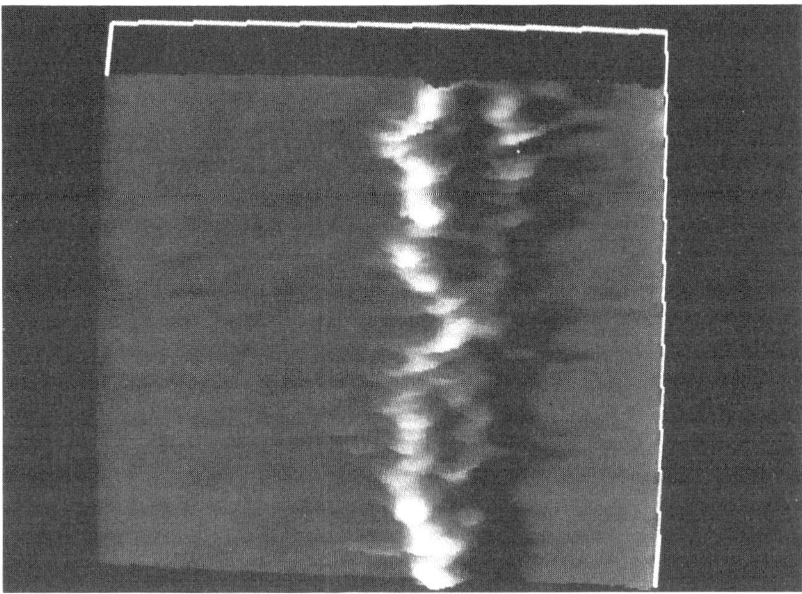

Fig. 13.22 AFM image of fibrinogen in the process of clotting. (Courtesy: B. Drake et al., University of California.)

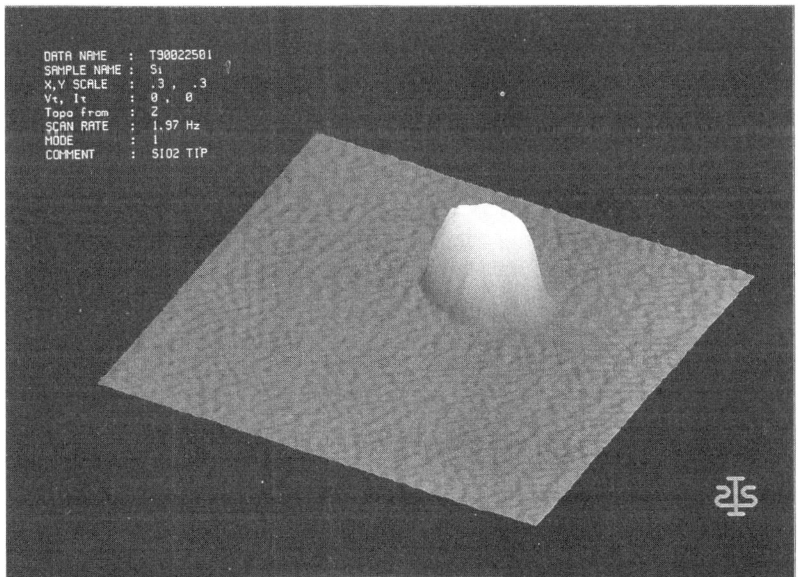

Fig. 13.23 AFM image of defects on silicon. (Courtesy: S. I. Park et al., Park Scientific Instruments.)

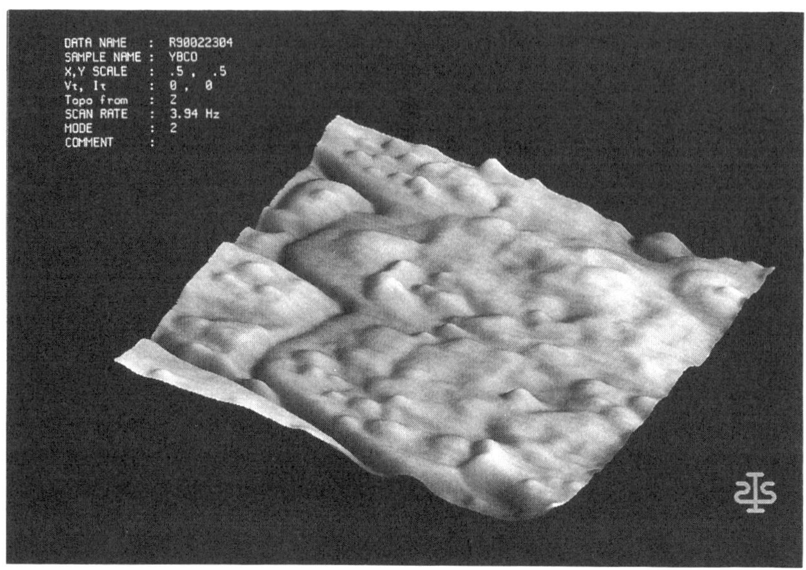

Fig. 13.24 AFM image of a superconductor. (Courtesy: S. I. Park, Park Scientific Instruments.)

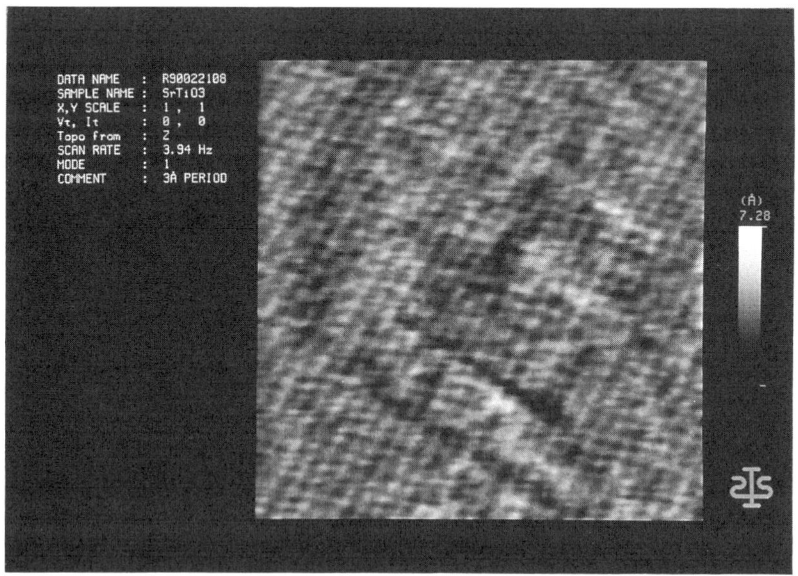

Fig. 13.25 AFM image of $SrTiO_3$. (Courtesy: S. I. Park, Park Scientific Instruments.)

Tables 13.2, 13.3, and 13.4 list references in which images of the surface of crystals, structures, biomaterials, and polymers have been reported. Table 13.5 lists references reporting images during the progress of processes. Tables 13.6 and 13.7, respectively, list references where surface modification has been performed and tribology characterized.

Table 13.2 Images of crystals.

Material	Reference
AgBr (100)	Heinzelmann et al. (1989a)
AgBr	Haefke et al. (1992)
Ag foil	Alexander et al. (1989)
Ag-doped CdTe	Bryant et al. (1988a)
AlGaAs superlattice	Chalmers et al. (1989)
Bi	Henriksen et al. (1990)
diamond film	Zimmermann-Edling et al. (1992)
Ge	Gould et al. (1990a)
gold on mica	Manne et al. (1990)
gold film	Taubenblatt (1989)
gold film	Schönenberger and Alvarado (1989)
HOPBN	Albrecht and Quate (1987)
HOPBN	Schmidt et al. (1990)
HOPG	Binnig et al. (1986, 1987)
HOPG	Albrecht and Quate (1987)
HOPG	Heinzelmann et al. (1987, 1989b)
HOPG	McClelland et al. (1987)
HOPG	Martin et al. (1987)
HOPG	Mate et al. (1987, 1989a)
HOPG	Bryant et al. (1988a)
HOPG	Meyer et al. (1988,1989a)
HOPG	Erlandsson et al. (1988a)
HOPG	Marti et al. (1988a,b,d)
HOPG in liquid	Schneir et al. (1988)
HOPG	Rugar et al. (1989)
HOPG	Gould et al. (1989a, 1990b)
HOPG	Burnham et al. (1989)
HOPG	Neubauer et al. (1990)
HOPG	Sugawara et al. (1990a,b)
Ir	Alexander et al. (1989)
LiF	Heinzelmann et al. (1988b, 1990)
LiF	Meyer et al. (1988, 1990a,c)

Table 13.2 (continued) Images of crystals.

Material	Reference
mica under solution	Drake et al. (1989)
mica	Schmidt et al. (1990)
molybdenite	Miller and Bryant (1989)
MoS_2	Albrecht and Quate (1987)
MoS_2 (He)	Kirk et al. (1988)
muscovite	Miller and Bryant (1989)
NaCl	Meyer and Amer (1988a,b)
NaCl (001)	Meyer and Amer (1990b)
Nb_3Sn	Anders and Heiden (1988, 1990)
PbS	Heinzelmann et al. (1990)
potassium bromide	Giessibl and Binnig (1992)
quartz	Heinzelmann et al. (1988b)
$RuCl_3$	Gould et al. (1990b)
sapphire	Barrett and Quate (1990)
selenite	Miller and Bryant (1989)
Si wafer	Martin et al. (1987)
Si wafer	Marti et al. (1988a,b)
native oxide of Si	Marti et al. (1988b)
native oxide of Si	Alexander et al. (1989)
Si (111)	Alexander et al. (1989)
Si	Sugawara et al. (1992)
high-T_c superconductor	Heinzelmann et al. (1988a)
$TaSe_2$, TaS_2	Garnaes and Hansma (1990)
$TaSe_2$, TaS_2	Meyer et al. (1990b)

Table 13.3 Images of structures.

Material	Reference
buckyball-substrate interactions	Sarid et al. (1992b)
Ag and Hg on Au(111)	Chen and Gewirth (1992)
C_{60}/C_{70} crystal surfaces	Dietz et al. (1992)
carbon coating	Meyer et al. (1989b)
compact disk	Anders and Heiden (1988, 1990)
fullerene films	Snyder et al. (1991)
magnetic recording head	Schönenberger et al. (1990)
magnetic sample	Schönenberger et al. (1990)
methylated SiO_2	Olsson et al. (1992)
optical flat	Heinzelmann et al. (1988b)
polished sapphire	Barrett and Quate (1990)
photographic film	Gould et al. (1990b)
photoresist	Martin et al. (1987)
Si integrated circuit	Gould et al. (1990a)
superlattice	Chalmers et al. (1989)
ultrafine clay particles	Garnaes et al. (1992)
zone plate	Meyer and Amer (1988a,b)

Table 13.4 Images of biomaterials and polymers.

Material	Reference
adenine and thymine	Allen et al. (1992a)
amino acid DL-leucine	Gould et al. (1988)
archaeobacterium	Mulhern et al. (1992)
aspirin	Masaki et al. (1992)
bacillus	Ohnesorge et al. (1992)
bacteria E. Coli	Gould et al. (1990a)
bacteriophages	Kolbe et al. (1992)
bacteriorhodopsin	Butt et al. (1990)
bacteriorhodopsin on purple membrane	Worcester et al. (1990)
cholesteric films	Moiseev et al. (1992)
dendritic crystals	Patil et al. (1990)
dialysis membranes	Kasper et al. (1992)

Table 13.4 (continued) Images of biomaterials and polymers.

Material	Reference
diatom	Linder et al. (1992)
didodecylbenzene	Eng et al. (1992a)
DMPE	Weisenhorn et al. (1991)
DMPG	Weisenhorn et al. (1991)
DNA	Gould et al. (1989b)
DNA	Weisenhorn et al. (1990b, 1991)
DNA	Prater et al. (1990)
DNA	Allen et al. (1992b)
DNA	Vesenka et al. (1992)
DNA	Thundat et al. (1992)
DNA	Heckl and Binnig (1992)
DODAB	Weisenhorn et al. (1991)
DPPG	Weisenhorn et al. (1991)
E. coli	Keller et al. (1992b)
fibrinogen	Gould et al. (1989b)
fibronectin	Emch et al. (1992)
layered organic crystals	Bosbach and Rammensee (1992)
leaf of house plant	Gould et al. (1990b)
lymphocite +IgG	Gould et al. (1990a)
Na,K-ATPase	Apell et al. (1992)
nitronyl nitroxide	Ruan et al. (1990)
organic conductors	Bar et al. (1992)
organic crystals	Overney et al. (1992)
organic molecular films	Yamada et al. (1992)
organic superconductors	Magonov et al. (1992)
perfluoro-poly propylene oxide	Mate et al. (1989b)
PMMA	Albrecht et al. (1988b)
PMMA	Dovek et al. (1988)
PMMA on Si	Williams et al. (1989)
PMPS	Dovek et al. (1988)
PODA	Albrecht et al. (1988b)
PODA	Dovek et al. (1988)
poly(1-butene)	Eng et al. (1992b)
polyalanine	Drake et al. (1989)
polyalanine	Gould et al. (1989b)
poly-L lysine	Gould et al. (1990a)

Table 13.4 (continued) Images of biomaterials and polymers.

Material	Reference
polymer	Hamada and Kaneko (1992)
polymerized monolayer of n-(2-aminoethyl)-10, 12-tricosadiynamide	Hansma et al. (1988), Marti et al. (1988c)
polymers	Stocker et al. (1992)
polyphenylalanine	Gould et al. (1989b)
polypropylene	Zajac et al. (1992)
protein	Schabert et al. (1992)
purple membrane	Worcester et al. (1988)
purple membrane	Gould et al. (1990a)
red blood cell	Gould et al. (1990a)
red and white blood cells	Hörber et al. (1990)
streptavidin	Weisenhorn et al. (1992)
tobacco mosaic	Zenhausern et al. (1992)
zeolite	Weisenhorn et al. (1990a)

Table 13.5. Processes.

Material	Reference
calcite crystal	Hillner et al. (1992)
clotting of human blood	Drake et al. (1989)
copper deposition on gold	Manne et al. (1991)
immunoglobulin IgG 4-4-20	Lin et al. (1990)
silicide lines	Robrock et al. (1989)
solid surfaces	Meyer et al. (1992)
stainless steel	Alexander et al. (1989)
underpotential deposition	Manne et al. (1991)

Table 13.6 Modification.

Material	Reference
gold films	Oden et al. (1992)
machining surfaces	Jung et al. (1992)
oxide thin films	Kim and Lieber (1992)
PMMA	Dovek et al. (1988)
PMPS	Dovek et al. (1988)
PODA	Albrecht et al. (1988b)
PODA	Dovek et al. (1988)
silicon surfaces	Lyo and Avouris (1991)
thermomechanical writing	Mamin and Rugar (1992)

Table 13.7 Indentation and tribology.

Material	Reference
Au-Ir	Cohen et al. (1990a)
Al_2O_3/Al	Burnham et al. (1989)
$CH_3(CH_2)_{16}$ $COOH/Al_2O_3/Al$	Burnham et al. (1989)
$CF_3(CH_2)_{16}$ $COOH/Al_2O_3/Al$	Burnham et al. (1989)
elastomer	Burnham and Colton (1989)
gold foil	Burnham and Colton (1989)
gold (111)	Cohen et al. (1990a)
gold film	Taubenblatt (1989)
HOPG	Mate et al. (1989b)
hydrogenated carbon coating	Heinzelmann et al. (1990)
layered organic crystals	Bosbach and Rammensee (1992)
lubricant	Mate et al. (1989b)
mica	Erlandsson et al. (1988b)
nanomechanics	Cohen (1992)
polymer lubricants	Mate (1992)
PTFE	Burnham et al. (1989)
Si and GaAs	Castell et al. (1992)
Xe and Kr	Krim and Chiarello (1990)

References

A

D. W. Abraham, C. C. Williams, and H. K. Wickramasinghe, "Measurement of in-plane magnetization by force microscopy," Appl. Phys. Lett. **53**, 1446 (1988a).

D. W. Abraham, C. C. Williams, and H. K. Wickramasinghe, "High-resolution force microscopy of in-plane magnetization," J. Micros. **152**, 863 (1988b).

D. W. Abraham, C. C. Williams, and H. K. Wickramasinghe, "Noise reduction technique for scanning tunneling microscopy," Appl. Phys. Lett. **53**, 1503 (1988c).

D. W. Abraham, C. C. Williams, and H. K. Wickramasinghe, "Differential scanning tunneling microscopy," J. Micros. **152**, 599 (1988d).

F. F. Abraham, I. P. Batra, and S. Ciraci, "Effect of tip profile on atomic-force microscope images: a model study," Phys. Rev. Lett. **60**, 1314 (1988e).

F. F. Abraham and I. P. Batra, "A theoretical interpretation of AFM images of graphite," Surf. Sci. **209**, L125 (1989).

D. W. Abraham and F. A. McDonald, "Theory of magnetic force microscope images," Appl. Phys. Lett. **56**, 1181 (1990).

G. P. Agrawal and N. K. Dutta, *Long-Wavelength Semiconductor Lasers* (Van Nostrand Reinhold, New York, 1986).

S. Akamine, R. C. Barrett, and C. F. Quate, "Improved atomic force microscope images using microcantilevers with sharp tips," Appl. Phys. Lett. (1990a).

S. Akamine, R. C. Barrett, and C. F. Quate, "Microfabrication of cantilevers with sharp tips for use in atomic force microscopy of large features," Proceedings of the STM'90/NANO I Conference, Baltimore, MD, July 23-27, 1990b.

T. R. Albrecht and C. F. Quate, "Atomic resolution imaging of a nonconductor by atomic force microscopy," J. Appl. Phys. **62**, 2599 (1987).

T. R. Albrecht and C. F. Quate, "Atomic resolution with the atomic force microscope on conductors and nonconductors," J. Vac. Sci. Technol. A **6**, 271 (1988).

T. R. Albrecht, M. M. Dovek, C. A. Lang, P. Grütter, C. F. Quate, S. W. J. Kuan, C. W. Frank, and R. F. W. Pease, "Imaging and modification of polymers by scanning tunneling and atomic force microscopy," J. Appl. Phys. **64**, 1178 (1988).

T. R. Albrecht, P. Grütter, D. Horne, and D. Rugar, "FM detection using high-Q cantilevers in vacuum for enhanced force microscope sensitivity," Proceedings of the STM'90/NANO I Conference, Baltimore, MD, July 23-27, 1990a.

T. R. Albrecht, S. Akamine, T. E. Carver, and C. F. Quate, "Microfabrication of cantilever styli for the atomic force microscope," J. Vac. Sci. Technol. A **8** (4), 3386 (1990b).

T. R. Albrecht, P. Grütter, D. Rugar, and D. P. E. Smith, "Low-temperature force microscope with all-fiber interferometer," Ultramicroscopy **42-44**, 1638 (1992).

S. Alexander, L. Hellemans, O. Marti, J. Schneir, V. Elings, P. K. Hansma, M. Longmire, and J. Gurley, "An atomic-resolution atomic-force microscope implemented using an optical lever," J. Appl. Phys. **65**, 164 (1989).

M. J. Allen, M. Balooch, S. Subbiah, R. J. Tench, R. Balhorn, and W. J. Siekhaus, "Analysis of adenine and thymine adsorbed on graphite by scanning tunneling and atomic force microscopy," Ultramicroscopy **42-44**, 1049 (1992a).

M. J. Allen, N. V. Hud, M. Balooch, R. J. Tench, W. J. Siekhaus, and R. Balhorn, "Tip-radius-induced artifacts in AFM images of protamine-complexed DNA fibers," Ultramicroscopy **42-44**, 1095 (1992b).

R. Allenspach, H. Salemink, A. Bischof, and E. Weibel, "Tunneling experiments involving magnetic tip and magnetic sample," Z. Phys. B **67**, 125 (1987).

R. Allenspach and A. Bischof, "Spin-polarized secondary electrons from a scanning tunneling microscope in field emission mode," Appl. Phys. Lett. **54**, 587 (1989).

M. Anders and C. Heiden, "Imaging of tip-sample compliance in STM," J. Micros. **152**, 643 (1988).

M. Anders, M. Mück, and C. Heiden, "Potentiometry for thin-film structures using atomic force microscopy," J. Vac. Sci. Technol. A **8**, 394 (1990).

M. Anders and C. Heiden, "Resolution in non-contact force microscopy," Proceedings of the STM'90/NANO I Conference, Baltimore, MD, July 23-27, 1990.

Y. Andoh, S. Oguchi, R. Kaneko, and T. Miyamoto, "Evaluations of lubricants by scanning tunneling microscope and frictional force microscope," Proceedings of the STM'90/NANO I Conference, Baltimore, MD, July 23-27, 1990.

D. Anselmetti, Ch. Gerber, B. Michel, H.-J. Güntherodt, and H. Rohrer, "Compact, combined scanning tunneling/force microscope," Rev. Sci. Instrum. **63**, 3003 (1992).

H.-J. Apell, J. Colchero, A. Linder, O. Marti, and J. Mlynek, "Na,K-ATPase in crystalline form investigated by scanning force microscopy," Ultramicroscopy **42-44**, 1133 (1992).

B

H. Bando, H. Tokumoto, W. Mizutani, K. Watanabe, M. Okano, M. Ono, H. Murakami, S. Okayama, Y. Ono, S. Wakiyama, F. Sakai, K. Endo, and K. Kajimura, "Effect of atomic force on the surface corrugation of 2H-$NbSe_2$ observed by scanning tunneling microscopy," Jpn. J. Appl. Phys. **26**, L41 (1987).

G. Bar, S. N. Magonov, H. J. Cantow, J. Gmeiner, and M. Schwoerer, "Atomic-scale imaging of anisotropic organic conductors by scanning probe techniques (STM/AFM)," Ultramicroscopy **42-44**, 644 (1992).

R. C. Barrett and C. F. Quate, "Imaging polished sapphire with atomic force microscopy," J. Vac. Sci. Technol. A **8**, 400 (1990).

R. C. Barrett and C. F. Quate, "Charge storage in a nitride-oxide-silicon medium by scanning capacitance microscopy," J. Appl. Phys. **70**, 2725 (1991).

R. C. Barrett and C. F. Quate, "Large-scale charge storage by scanning capacitance microscopy," Ultramicroscopy **42-44**, 262 (1992).

L. Bartolini, G. Fornetti, M. Ferri De Collibus, G. Occhionero, and F. Papetti, "Two-wavelength infrared heterodyne transceiver with a continuous phase tracking system," Rev. Sci. Instrum. **61**, 1177 (1990).

F. Bergasa and J. J. Sáenz, "Is it possible to observe biological macromolecules by electrostatic force microscopy?" Ultramicroscopy **42-44**, 1189 (1992).

G. Binnig and H. Rohrer, "Scanning tunneling microscopy," Helv. Phys. Acta **55**, 726 (1982).

G. Binnig and H. Rohrer, "Scanning tunneling microscopy," IBM J. Res. Dev. **30**, 355 (1986).

G. Binnig, C. F. Quate, and Ch. Gerber, "Atomic force microscope," Phys. Rev. Lett. **56**, 930 (1986).

G. Binnig, Ch. Gerber, E. Stoll, T. R. Albrecht, and C. F. Quate, "Atomic resolution with atomic force microscope," Europhys. Lett. **3**, 1281 (1987).

G. Binnig, "Force microscopy," Ultramicroscopy **42-44**, 7 (1992).

G. S. Blackman, C. M. Mate, M. R. Philpott, and G. A. Somorjai, "Studies of molecular wear with an atomic force microscope," Proceedings of the STM'90/NANO I Conference, Baltimore, MD, July 23-27, 1990.

F. R. Blom, S. Bouwstra, M. Elwenspoek, J. H. J. Fluitman, "Dependence of the quality factor of micromachined silicon beam resonators on pressure and geometry," J. Vac. Sci. Technol. B **10**, 19 (1992).

D. Bosbach and W. Rammensee, "Surface manipulation on layered organic crystals by scanning force microscopy," Ultramicroscopy **42-44**, 973 (1992).

X. Bouju, C. Girard, and B. Labani, "Self-consistent study of the electromagnetic coupling between a thin probe tip and a surface: implication for atomic-force and near-field microscopy," Ultramicroscopy **42-44**, 430 (1992).

G. Bozzolo and J. Ferrante, "Theoretical modelling of AFM for bimetallic tip-substrate interactions," Ultramicroscopy **42-44**, (1992).

S. Breen, B. E. Paton, B. L. Blackford, and M. H. Jericho, "Fiber optic displacement sensor with subangstrom resolution," Appl. Opt. **29**, 16 (1990).

G. E. Bridges and D. J. Thomson, "High-frequency circuit characterization using the AFM as a reactive near-field probe," Ultramicroscopy **42-44**, 321 (1992).

P. J. Bryant, R. G. Miller, R. Deeken, R. Yang, and Y. C. Zheng, "Scanning tunneling and atomic force microscopy performed with the same probe in one unit," J. Micros. **152**, 871 (1988a).

P. J. Bryant, R. G. Miller, and R. Yang, "Scanning tunneling and atomic force microscopy combined," Appl. Phys. Lett. **52**, 2233 (1988b).

P. J. Bryant, R. G. Miller, R. H. Deeken, R. Yang, and Y. C. Zheng, "Scanning tunneling microscopy and atomic force microscopy performed with the same probe in one unit," J. Micros. **152**, 871 (1989).

P. J. Bryant, Y. C. Cheng, and H. S. Kim, "Tunnel junction atomic force microscopy," Proceedings of the STM'90/NANO I Conference, Baltimore, MD, July 23-27, 1990.

D. Bürgler, H. Heinzelmann, G. Tarrach, E. Meyer, and H.-J. Güntherodt, "Atomic force microscope for the ultra-high vacuum," Proceedings of the STM'90/NANO I Conference, Baltimore, MD, July 23-27, 1990.

N. A. Burnham and R. J. Colton, "Measuring the nanomechanical properties and surface forces of materials using an atomic force microscope," J. Vac. Sci. Technol. A **7**, 2906 (1989).

N. A. Burnham, D. D. Dominguez, R. L. Mowery, and R. J. Colton, "Probing the surface forces of monolayer films with an atomic force microscope," Washington, D.C., Naval Research Laboratory, preprint 1989.

N. A. Burnham, H. M. Pollock, W. B. Barger, R. J. Colton, I. P. Hayward, D. S. Y. Hsu, and I. L. Singer, "Nanometer-scale surface mechanical properties and imaging of diamond," Proceedings of the STM'90/NANO I Conference, Baltimore, MD, July 23-27, 1990.

R. A. Buser, J. Brugger, and N. F. de Rooij, "Micromachined silicon cantilevers and tips for bidirectional force microscopy," Ultramicroscopy **42-44**, 1476 (1992).

H.-J. Butt, C. B. Prater, and P. K. Hansma, "Imaging purple membrane dry and in water with the atomic force microscope," Proceedings of the STM'90/NANO I Conference, Baltimore, MD, July 23-27, 1990.

C

M. R. Castell, M. G. Walls, and A. Howie, "Imaging of low-load indentations into Si and GaAs by scanning tunneling microscopy," Ultramicroscopy **42-44**, 1490 (1992).

S. A. Chalmers, A. C. Gossard, A. L. Weisenhorn, S. A. C. Gould, B. Drake, and P. K. Hansma, "Determination of tilted superlattice structure by atomic force microscopy," Appl. Phys. Lett. **55**, 2491 (1989).

C.-H. Chen and A. A. Gewirth, "AFM study of the structure of underpotentially deposited Ag and Hg on Au(111)," Ultramicroscopy **42-44**, 437 (1992).

C. Julian Chen, "Attractive interatomic force as a tunnelling phenomenon," J. Phys. Cond. Matter **3**, 1227 (1991).

C. Julian Chen and R. J. Hamers, "Role of atomic force in tunneling-barrier measurements," J. Vac. Sci. Technol. B **9**, 503 (1991).

C. Julian Chen, "In situ testing and calibration of tube piezoelectric scanners," Ultramicroscopy **42-44**, 1653 (1992a).

C. Julian Chen, "Electromechanical deflections of piezoelectric tubes with quartered electrodes," Appl. Phys. Lett. **60**, 132 (1992b).

C. Julian Chen, *Introduction to Scanning Tunneling Microscopy* (Oxford University Press, New York, 1993).

Y. C. Cheng, H. S. Kim, R. H. Deeken, and P. J. Bryant, "Tunnel junction atomic force microscopy," Proceedings of the STM'90/NANO I Conference, Baltimore, MD, July 23-27, 1990.

S. Ciraci, "Atomic-scale tip-sample interactions and contact phenomena," Ultramicroscopy **42-44**, 16 (1992).

S. Ciraci, E. Tekman, M. Gökcedag, I. P. Batra, and A. Baratoff, "Adhesive energy, force and barrier height between simple metal surfaces," Ultramicroscopy **42-44**, 163 (1992).

S. R. Cohen, G. Neubauer, and G. M. McClelland, "Nanomechanics of a Au-Ir contact using a bidirectional atomic force microscope," J. Vac. Sci. Technol. **8** (4), 3449 (1990a).

S. R. Cohen, G. Neubauer, G. M. McClelland, D. Coulman, and C. A. Brown, "Nanotribology using a bidirectional AFM," Proceedings of the STM'90/NANO I Conference, Baltimore, MD, July 23-27, 1990b.

S. R. Cohen, "An evaluation of the use of the atomic force microscope for studies in nanomechanics," Ultramicroscopy **42-44**, 66 (1992).

J. H. Coombs and J. B. Pethica, "Properties of vacuum tunneling currents: anomalous barrier heights," IBM J. Res. Dev. **30**, 455 (1986).

D

A. Dandridge, A. B. Tveten, R. O. Miles, and T. G. Giallorenzi, "Laser noise in fiber-optic interferometer systems," Appl. Phys. Lett. **37**, 526 (1980a).

A. Dandridge, R. O. Miles, and T. G. Giallorenzi, "Diode laser sensor," Electron. Lett. **16**, 948 (1980b).

A. Dandridge, A. B. Tveten, R. O. Miles, D. A. Jackson, and T. G. Giallorenzi, "Single-mode diode laser phase noise," Appl. Phys. Lett. **38**, 77 (1981).

M. S. Daw and M. I. Baskes, "Semiempirical, quantum mechanical calculation of hydrogen embrittlement in metals," Phys. Rev. Lett. **50**, 1285 (1983).

M. S. Daw and M. I. Baskes, "Embedded-atom method: derivation and application to impurities, surfaces, and other defects in metals," Phys. Rev. B **29**, 6443 (1984).

A. J. den Boef, "Scanning force microscopy using a simple low-noise interferometer," Appl. Phys. Lett. **55**, 439 (1989).

A. J. den Boef, "Preparation of magnetic tips for a scanning force microscope," Appl. Phys. Lett. **56**, 2045 (1990).

A. J. den Boef, "Scanning force microscopy using optical interferometry," Ph.D. thesis, Philips Research Laboratories, 1991.

B. V. Derjaguin, V. M. Muller, and Yu. P. Toporov, "Effect of contact deformations on the adhesion of particles," J. Coll. Interface Sci. **53**, 314 (1975).

W. Denk and D. W. Pohl, "Local electrical dissipation imaged by scanning force microscopy," Appl. Phys. Lett. **59**, 2171 (1991).

A. DiCarlo, M. R. Scheinfein, and R. V. Chamberlin, "Magnetic force microscope utilizing and ultra-small spring constant vertically cantilevered tip," Ultramicroscopy **42–44**, 383 (1992).

P. Dietz, K. Fostiropoulos, W. Krätschmer, and P. K. Hansma, "Size and packing of fullerenes on C_{60}/C_{70} crystal surfaces studied by atomic force microscopy," Appl. Phys. Lett. **60**, 62 (1992).

Digital Instruments Inc. (Nanoscope II), 6780 Cortona Drive, Santa Barbara, CA. 93117.

M. M. Dovek, T. R. Albrecht, S. W. J. Kuan, C. A. Lang, R. Emch, P. Grütter, C. W. Frank, R. F. W. Pease, and C. F. Quate, "Observation and manipulation of polymers by scanning tunneling and atomic force microscopy," J. Micros. **152**, 229 (1988).

B. Drake, C. B. Prater, A. L. Weisenhorn, S. A. C. Gould, T. R. Albrecht, C. F. Quate, D. S. Cannell, H. G. Hansma, and P. K. Hansma, "Imaging crystals, polymers, and processes in water with the atomic force microscope," Science **243**, 1586 (1989).

K. E. Drexler, "Molecular tip arrays for AFM imaging and nanofabrication," Proceedings of the STM'90/NANO I Conference, Baltimore, MD, July 23-27, 1990.

U. Dürig, J. K. Gimzewski, and D. W. Pohl, "Experimental observation of forces acting during scanning tunneling microscopy," Phys. Rev. Lett. **57**, 2403 (1986).

U. Dürig, O. Züger, and D. W. Pohl, "Force sensing in scanning tunneling microscopy: observation of adhesion forces on clean metal surfaces," J. Micros. **152**, 259 (1988).

U. Dürig and O. Züger, "Study of metallic adhesion using scanning tunneling microscopy," Proceedings of the International Vacuum Congress (IVC-11) Cologne, W. Germany, September 25-29, 1989.

U. Dürig, O. Züger, and A. Stalder, "Interaction force detection in scanning probe microscopy: methods and applications," J. Appl. Phys. **72**, 1778 (1992).

E

R. Emch, F. Zenhausern, M. Jobin, M. Taborelli, and P. Descouts, "Morphological difference between fibronectin sprayed on mica and on PMMA," Ultramicroscopy **42-44**, 1155 (1992).

L. M. Eng, H. Fuchs, S. Buchholz, and J. P. Rabe, "Ordering of didodecylbenzene on graphite: a combined SFM/STM study," Ultramicroscopy **42-44**, 1059 (1992a).

L. M. Eng, H. Fuchs, K. D. Jandt, and J. Petermann, "Investigating poly(1-butene) films by SFM/STM," Ultramicroscopy **42-44**, 989 (1992b).

R. Erlandsson, G. M. McClelland, C. M. Mate, and S. Chiang, "Atomic force microscopy using optical interferometry," J. Vac. Sci. Technol. A **6**, 266 (1988a).

R. Erlandsson, G. Hadziioannou, C. M. Mate, G. M. McClelland, and S. Chiang, "Atomic scale friction between the muscovite mica cleavage plane and a tungsten tip," J. Chem Phys. **89**, 5190 (1988b).

F

S. M. Foiles, M. I. Baskes, and M. S. Daw, "Embedded-atom-method functions for the fcc metals Cu, Ag, Au, Ni, Pd, Pt, and their alloys," Phys. Rev. B **33**, 7983 (1986).

G

J. Garnaes and P. K. Hansma, "Atomic force microscopy of charge-density waves and atoms on 1T-$TaSe_2$, 1T-TaS_2, 1T-$TiSe_2$ and 2H-$NbSe_2$," Proceedings of the STM'90/NANO I Conference, Baltimore, MD, July 23-27, 1990.

J. Garnaes, H. Lindgreen, P. L. Hansen, S. A. C. Gould, and P. K. Hansma, "Atomic force microscopy of ultrafine clay particles," Ultramicroscopy **42-44**, 1428 (1992).

F. J. Giessibl, D. P. E. Smith, Ch. Gerber, and G. Binnig, "A low temperature atomic force/scanning tunneling microscope for ultrahigh vacuum," Proceedings of the STM'90/NANO I Conference, Baltimore, MD, July 23-27, 1990.

F. J. Giessibl, "Theory for an electrostatic imaging mechanism allowing atomic resolution of ionic crystals by atomic force microscopy," Phys. Rev. B **45**, 13815 (1992).

F. J. Giessibl and G. Binnig, "Investigation of the (001) cleavage plane of potassium bromide with an atomic force microscope at 4.2 K in ultra-high vacuum," Ultramicroscopy **42-44**, 281 (1992).

J. K. Gimzewski and R. Möller, "Transition from the tunneling regime to point contact studied using scanning tunneling microscopy," Phys. Rev. B **36**, 1284 (1987).

J. K. Gimzewski, R. Möller, D. W. Pohl, and R. R. Schlittler, "Transition from tunneling to point contact investigated by scanning tunneling microscopy and spectroscopy," Surf. Sci. **189-190**, 15 (1987).

C. Girard, "Theoretical atomic-force-microscopy study of a stepped surface: nonlocal effects in the probe," Phys. Rev. B **43**, 8822 (1991).

P. Gleyzes, P. K. Kuo, and A. C. Boccara, "Bistable behavior of a vibrating tip near a solid surface," Appl. Phys. Lett. **58**, 2989 (1991).

T. Göddenhenrich, U. Hartmann, M. Anders, and C. Heiden, "Investigation of Bloch wall fine structure by magnetic force microscopy," J. Micros. **152**, 527 (1988).

T. Göddenhenrich, H. Lemke, U. Hartmann, and C. Heiden, "Force microscope with capacitive displacement detection," J. Vac. Sci. Technol. A **8**, 383 (1990).

T. Göddenhenrich, U. Hartmann, and C. Heiden, "Generation and imaging of domains with the magnetic force microscope," Ultramicroscopy **42-44**, 256 (1992).

Frank O. Goodman and Nicolas Garcia, "Roles of the attractive and repulsive forces in atomic-force microscopy," Phys. Rev. B **43**, 4728 (1991).

S. Gould, O. Marti, B. Drake L. Hellemans, C. E. Bracker, P. K. Hansma, N. L. Keder, M. M. Eddy, and G. D. Stucky, "Molecular resolution images of amino acid crystals with the atomic force microscope," Nature **332**, 332 (1988).

S. A. C. Gould, K. Burke, and P. K. Hansma, "Simple theory for the atomic-force microscope with a comparison of theoretical and experimental images of graphite," Phys. Rev. B **40**, 5363 (1989a).

S. A. C. Gould, B. Drake, C. B. Prater, A. L. Weisenhorn, S. M. Lindsay, and P. K. Hansma, "Imaging polymers, proteins and DNA in aqueous solutions with the atomic force microscope," Proceedings of the 47th Annual Meeting of the Electron Microscopy Society of America (1989b).

S. A. C. Gould, B. Drake, C. B. Prater, A. L. Weisenhorn, S. Manne, H. G. Hansma, P. K. Hansma, J. Massie, M. Longmire, V. Elings, B. Dixon Northern, B. Mukergee, C. M. Peterson, W. Stoeckenius, T. R. Albrecht, and C. F. Quate, "From atoms to integrated circuit chips, blood cells, and bacteria with the atomic force microscope," J. Vac. Sci. Technol. A **8**, 369 (1990a).

S. A. C. Gould, B. Drake, C. B. Prater, A. L. Weisenhorn, S. Manne, G. L. Kelderman, H.-J. Butt, H. Hansma, P. K. Hansma, and S. Maganov, "The atomic force microscope: a tool for science and industry," (1990b).

S. A. C. Gould, H.-J. Butt, and P. K. Hansma, "Tip-surface interaction in the atomic force microscope: experiment and theory," Proceedings of the STM'90/NANO I Conference, Baltimore, MD, July 23-27, 1990c.

D. A. Grigg, P. E. Russell, J. E. Griffith, M. J. Vasile, and E. A. Fitzgerald, "Probe characterization for scanning probe metrology," Ultramicroscopy **42-44**, 1616 (1992a).

D. A. Grigg, P. E. Russell, and J. E. Griffith, "Rocking-beam force-balance approach to atomic force microscopy," Ultramicroscopy **42-44**, 1504 (1992b).

P. Grütter, E. Meyer, H. Heinzelmann, L. Rosenthaler, H.-R. Hidber, and H.-J. Güntherodt, "Application of atomic force microscopy to magnetic materials," J. Vac. Sci. Technol. A **6**, 279 (1988).

P. Grütter, A. Wadas, E. Meyer, H.-R. Hidber, and H.-J. Güntherodt, "Magnetic force microscopy of a CoCr thin film," J. Appl. Phys. **66**, 6001 (1989).

P. Grütter, A. Wadas, E. Meyer, H. Heinzelmann, H.-R. Hidber, and H.-J. Güntherodt, "High resolution magnetic force microscopy," J. Vac. Sci. Technol. A **8**, 406 (1990a).

P. Grütter, Th. Jung, H. Heinzelmann, A. Wadas, E. Meyer, H.-R. Hidber, and H.-J. Güntherodt, "10 nm resolution by magnetic force microscopy on FeNdB," J. Appl. Phys. **67**, 1437 (1990b).

P. Grütter, D. Rugar, H. J. Mamin, G. Castello, C.-J. Lin, and B. Valletta, "Batch fabricated cantilevers for magnetic force microscopy," Proceedings of the STM'90/NANO I Conference, Baltimore, MD, July 23-27, 1990c.

P. Grütter, H.-J. Mamin, and D. Rugar, in *Scanning Tunneling Microscopy II*, edited by R. Wiesendanger and H.-J. Güntherodt (Springer-Verlag, Berlin, 1992) pp. 151-207.

H

H. Haefke, E. Meyer, L. Howald, U. D. Schwarz, G. Gerth, and M. Krohn, "Atomic surface and lattice structures of AgBr thin films," Ultramicroscopy **42-44**, 290 (1992).

E. Hamada and R. Kaneko, "Micro-tribological evaluations of a polymer surface by atomic force microscopes," Ultramicroscopy **42-44**, 184 (1992).

K. Hane, S. Watanabe, and T. Goto, "Force microscope using a twin-path interferometer," Proceedings of the STM'90/NANO I Conference, Baltimore, MD, July 23-27, 1990.

P. K. Hansma, V. B. Elings, O. Marti, and C. E. Bracker, "Scanning tunneling microscopy and atomic force microscopy: application to biology and technology," Science **242**, 209 (1988).

H. W. Hao, A. M. Baró, and J. J. Sáenz, "Electrostatic and capillary forces in force microscopy," Proceedings of the STM'90/NANO I Conference, Baltimore, MD, July 23-27, 1990.

C. M. Harris ed., *Shock and Vibration Handbook* (McGraw-Hill Book Company, New York, 1988).

W. A. Harrison, "Total energies in the tight-binding theory," Phys. Rev. B **23**, 5230 (1981).

U. Hartmann and H. H. Mende, "Observation of Bloch wall fine structures on iron whiskers by a high-resolution interference contrast technique," J. Phys. D: Appl. Phys. **18**, 2285 (1985).

U. Hartmann, "Magnetic force microscopy: some remarks from the micromagnetic point of view," J. Appl. Phys. **64**, 1561 (1988).

U. Hartmann and C. Heiden, "Calculation of the Bloch wall contrast in magnetic force microscopy," J. Micros. **152**, 281 (1988).

U. Hartmann, "Theory of magnetic force microscopy," J. Vac. Sci. Technol. A **8**, 411 (1990a).

U. Hartmann, "Theory of van der Waals microscopy," Proceedings of the STM'90/NANO I Conference, Baltimore, MD, July 23-27, 1990b.

U. Hartmann, "Intermolecular and surface forces in noncontact scanning force microscopy," Ultramicroscopy **42-44**, 59 (1992).

W. M. Heckl, M. Sprecht, F. Ohnesorge, and M. Hashmi, "Ring structure on natural molybdenum disulfide investigated by scanning tunneling and scanning force microscopy," Proceedings of the STM'90/NANO I Conference, Baltimore, MD, July 23-27, 1990.

W. M. Heckl and G. Binnig, "Domain walls on graphite mimic DNA," Ultramicroscopy **42-44**, 1073 (1992).

C. V. Heer, *Statistical Mechanics, Kinetic Theory and Stochastic Processes* (Academic Press, New York, 1972).

H. Heinzelmann, P. Grütter, E. Meyer, H. Hidber, L. Rosenthaler, M. Ringger, and H.-J. Güntherodt, "Design of an atomic force microscope and first results," Surf. Sci. **189/190**, 29 (1987).

H. Heinzelmann, D. Anselmetti, R. Wiesendanger, H.-R. Hidber, H.-J. Güntherodt, M. Düggelin, R. Guggenheim, H. Schmidt, and G. Güntherodt, "STM and AFM investigations of high-T_c superconductors," J. Micros. **152**, 399 (1988a).

H. Heinzelmann, E. Meyer, P. Grütter, H.-R. Hidber, L. Rosenthaler, and H.-J. Güntherodt, "Atomic force microscopy - general aspects and application to insulators," J. Vac. Sci. Technol. A **6**, 275 (1988b).

H. Heinzelmann, E. Meyer, L. Scandella, P. Grütter, Th. Jung, H. Hug, H.-R. Hidber, and H.-J. Güntherodt, "Topography and correlation to wear of hydrogenated amorphous carbon coatings: an atomic force microscopy study," *Wear*, edited by F. F. Ling and C. H. T. Pan (Springer-Verlag, New York, 1988c).

H. Heinzelmann, E. Meyer, H.-J. Güntherodt, and R. Steiger, "Local step structure of the AgBr(100) and (111) surfaces studied by atomic force microscopy," Surf. Sci. **221**, 1 (1989a).

H. Heinzelmann, E. Meyer, H. Rudin, and H.-J. Güntherodt, "Force microscopy," Proceedings of the NATO Meeting "Basic Concepts and Applications of Scanning Tunneling Microscopy (STM) and Related Techniques," Erice, Sicily, April 17-29, 1989b.

H. Heinzelmann, E. Meyer, D. Brodbeck, G. Overney, and H.-J. Güntherodt, "Atomic resolution imaging on layered and non-layered materials by AFM," Proceedings of the STM'90/NANO I Conference, Baltimore, MD, July 23-27, 1990.

L. Hellemans, K. Waeyaert, and F. Hennau, "Can AFM tips be inspected by atomic force microscopy?" Proceedings of the STM'90/NANO I Conference, Baltimore, MD, July 23-27, 1990.

P. N. Henriksen, H. T. Chu, and R. D. Ramsier, "Atomically resolved images of Bismuth films on mica with atomic force microscope," Proceedings of the STM'90/NANO I Conference, Baltimore, MD, July 23-27, 1990.

P. E. Hillner, S. Manne, A. J. Gratz, and P. K. Hansma, "AFM images of dissolution and growth on a calcite crystal," Ultramicroscopy **42-44**, 1387 (1992).

M. Hipp, H. Bielefeldt, J. Colchero, O. Marti, and J. Mlynek, "A stand-alone scanning force and friction microscope," Ultramicroscopy **42-44**, 1498 (1992).

P. C. D. Hobbs, D. W. Abraham, and H. K. Wickramasinghe, "Magnetic force microscopy with 25 nm resolution," Appl. Phys. Lett. **55**, 2357 (1989).

J. K. H. Hörber, W. Häberle, and G. Binnig, "Force microscopy on living cells," Proceedings of the STM'90/NANO I Conference, Baltimore, MD, July 23-27, 1990.

S. Hosaka, Y. Honda, S. Hosoki, and T. Hasegawa, "Surface observations of insulator and magnetic materials using a prototype AFM and MFM," Proceedings of the STM'90/NANO I Conference, Baltimore, MD, July 23-27, 1990.

Y. J. Huang, J. Slinkman, and C. C. Williams, "Modelling of impurity dopant density measurement in semiconductors by scanning force microscopy," Ultramicroscopy **42-44**, 298 (1992).

H. Hug, Th. Jung, T. Frey, D. Anselmetti, D. Brodbeck, and H.-J. Güntherodt, "Experiments with the magnetic force microscope on high temperature superconductors: theoretical aspects and first results," Proceedings of the STM'90/NANO I Conference, Baltimore, MD, July 23-27, 1990.

I

T. Iijima and K. Yasuda, "Submicron-scale tip fabrication for magnetic force microscopy by electrolytic polishing," Jpn. J. Appl. Phys. **27**, 1546 (1988).

J. N. Israelachvili, "The calculation of van der Waals dispersion forces between macroscopic bodies," Proc. R. Soc. London, Ser. A **331**, 39 (1972).

J. N. Israelachvili and D. Tabor, "The measurement of van der Waals dispersion forces in the range of 1.5 to 130 nm," Proc. R. Soc. London, Ser. A **331**, 19 (1972).

J. N. Israelachvili, *Intermolecular and Surface Forces* (Academic Press, New York, 1985).

J. N. Israelachvili, P. M. McGuiggan, and A. M. Homola, "Dynamic properties of molecularly thin liquid films," Science **240**, 189 (1988).

J. N. Israelachvili, *Intermolecular and Surface Forces*, 2nd ed. (Academic Press, New York, 1991).

J

T. Jung, H. Hug, E. Meyer, D. Brodbeck, H. R. Hidber, and H.-J. Güntherodt, "AFM and STM on microfabricated submicrometer structures: imaging problems in large scale measurements," Proceedings of the STM'90/NANO I Conference, Baltimore, MD, July 23-27, 1990.

T. A. Jung, A. Moser, H. J. Hug, D. Brodbeck, R. Hofer, H. R. Hidber, and U. D. Schwarz, "The atomic force microscope used as a powerful tool for machining surfaces," Ultramicroscopy **42-44**, 1446 (1992).

K

H. Kado, K. Yokoyama, and T. Tohda, "A novel ZnO whisker tip for atomic force microscopy," Ultramicroscopy **42-44**, 1659 (1992).

K. Kasper, K.-H. Herrmann, P. Dietz, P. K. Hansma, O. Inacker, H.-D. Lehmann, and Th. Rintelen, "Investigation of dialysis membranes with atomic force microscopy," Ultramicroscopy **42-44**, 1181 (1992).

D. Keller, D. Deputy, A. Alduino, and K. Luo, "Sharp, vertical-walled tips for SFM imaging of steep or soft samples," Ultramicroscopy **42-44**, 1481 (1992a).

R. W. Keller, D. J. Keller, D. Bear, J. Vesenka, and C. Bustamante, "Electrodeposition procedure of *E. coli* RNA polymerase onto gold and deposition of *E. coli* RNA polymerase onto mica for observation with scanning force microscopy," Ultramicroscopy **42-44**, 1173 (1992b).

Y. Kim and C. M. Lieber, "Machining oxide thin films with an atomic force microscope: pattern and object formation on the nanometer scale," Science **257**, 375 (1992).

M. D. Kirk, T. R. Albrecht, and C. F. Quate, "Low-temperature atomic force microscopy," Rev. Sci. Instrum. **59**, 833 (1988).

W. F. Kolbe, D. F. Ogletree, and M. B. Salmeron, "Atomic force microscopy imaging of T4 bacteriophages on silicon substrates," Ultramicroscopy **42-44**, 1113 (1992).

L. C. Kong, B. G. Orr, K. D. Wise, C. Orme, and J. Sudijono, "A silicon micromachined sensor for force microscopy," Proceedings of the STM'90/NANO I Conference, Baltimore, MD, July 23-27, 1990.

J. Krim and R. Chiarello, "Sliding friction measurements of physisorbed monolayers: a comparison of solid and liquid films," Proceedings of the STM'90/NANO I Conference, Baltimore, MD, July 23-27, 1990.

Y. Kuk and P. J. Silverman, "Scanning tunneling microscope instrumentation," Rev. Sci. Instrum. **60**, 165 (1989).

L

Uzi Landman, W. D. Luedtke, and A. Nitzan, "Dynamics of tip-substrate interactions in atomic force microscopy," Surf. Sci. **210**, L177 (1989a).

Uzi Landman, W. D. Luedtke, and M. W. Ribarsky, "Structural and dynamical consequences of interactions in interfacial systems," J. Vac. Sci. Technol. A **7**, 2829 (1989b).

Uzi Landman, W. D. Luedtke, and M. W. Ribarsky, "Micromechanics and microdynamics via atomistic simulations," in *New Materials Approaches to Tribology: Theory and Applications*, edited by L. E. Pope, L. Fehrenbacher, and W. O. Winer (MRS, Boston, 1989c).

Uzi Landman, W. D. Luedtke, N. A. Burnham, and R. J. Colton, "Atomistic mechanisms and dynamics of adhesion, nanoindentation, and fracture," Science **248**, 454 (1990).

Uzi Landman and W. D. Luedtke, "Nanomechanics and dynamics of tip-substrate interactions," Proceedings of the STM'90/NANO I Conference, Baltimore, MD, July 23-27, 1990.

J. N. Lin, B. Drake, A. S. Lea, P. K. Hansma, and J. D. Andrade, "Direct observation of immunoglobulin adsorption dynamics using the atomic force microscope," Langmuir **6**, 509 (1990).

A. Linder, J. Colchero, H.-J. Apell, O. Marti, and J. Mlynek, "Scanning force microscopy of diatom shells," Ultramicroscopy **42-44**, 329 (1992).

I.-W. Lyo and P. Avouris, "Field-induced nanometer- to atomic-scale manipulation of silicon surfaces with the STM," Science Reports, July 1991, p. 173.

M

S. N. Magonov, G. Bar, E. Keller, E. B. Yagubskii, E. E. Laukhina, and H.-J. Cantow, "Surface analysis of organic superconductors by scanning probe techniques: STM and AFM," Ultramicroscopy **42-44**, 1009 (1992).

H. J. Mamin, D. Rugar, J. E. Stern, B. D. Terris, and S. E. Lambert, "Force microscopy of magnetization patterns in longitudinal recording media," Appl. Phys. Lett. **53**, 1563 (1988).

H. J. Mamin, D. Rugar, J. E. Stern, R. E. Fonatana, Jr., and P. Kasiraj, "Magnetic force microscopy of thin Permalloy films," Appl. Phys. Lett. **55**, 318 (1989).

H. J. Mamin and D. Rugar, "Thermomechanical writing with an atomic force microscope tip," Appl. Phys. Lett. **61**, 1003 (1992).

S. Manne, H. J. Butt, S. A. C. Gould, and P. K. Hansma, "Imaging metal atoms in air and water using the atomic force microscope," Appl. Phys. Lett. **56**, 1758 (1990).

S. Manne, P. K. Hansma, J. Massie, V. B. Elings, A. A. Gewirth, "Atomic-resolution electrochemistry with the atomic force microscope: copper deposition on gold," Science Reports, Jan. 1991, p. 183.

M. Mansuripur, "Computation of fields and forces in magnetic force microscopy," IEEE Trans. Magn. **25**, 3467 (1989a).

M. Mansuripur, "Demagnetizing field computation for thin films: extension to the hexagonal lattice," J. Appl. Phys. **66**, 3731 (1989b).

O. Marti, B. Drake, and P. K. Hansma, "Atomic force microscopy of liquid-covered surfaces: atomic resolution images," Appl. Phys. Lett. **51**, 484 (1987).

O. Marti, B. Drake, S. Gould, and P. K. Hansma, "Atomic force microscopy and scanning tunneling microscopy with a combination atomic force microscope/scanning tunneling microscope," J. Vac. Sci. Technol. A **6**, 2089 (1988a).

O. Marti, B. Drake, S. Gould, and P. K. Hansma, "Atomic resolution atomic force microscopy of graphite and the 'native oxide' on silicon," J. Vac. Sci. Technol. A **6**, 287 (1988b).

O. Marti, H. O. Ribi, B. Drake, T. R. Albrecht, C. F. Quate, and P. K. Hansma, "Atomic force microscopy of an organic monolayer," Science **239**, 50 (1988c).

O. Marti, B. Drake, S. Gould, and P. K. Hansma, "Probing surfaces with the atomic force microscope," SPIE **897**, *Scanning Microscopy Technologies and Applications* (1988d).

O. Marti, S. Gould, and P. K. Hansma, "Control electronics for atomic force microscopy," Rev. Sci. Instrum. **59**, 836 (1988e).

O. Marti, J. Colchero, and J. Mlynek, "Combined scanning force and friction microscopy of mica," Nanotechnology **1**, 141 (1990).

Y. Martin, C. C. Williams, and H. K. Wickramasinghe, "Atomic force microscope-force mapping and profiling on a sub 100-Å scale," J. Appl. Phys. **61**, 4723 (1987).

Y. Martin and H. K. Wickramasinghe, "Magnetic imaging by 'force microscopy' with 1000 Å resolution," Appl. Phys. Lett. **50**, 1455 (1987).

Y. Martin, D. Rugar, and H. K. Wickramasinghe, "High-resolution magnetic imaging of domains in TbFe by force microscopy," Appl. Phys. Lett. **52**, 244 (1988a).

Y. Martin, D. W. Abraham, and H. K. Wickramasinghe, "High-resolution capacitance measurement and potentiometry by force microscopy," Appl. Phys. Lett. **52**, 1103 (1988b).

N. Masaki, K. Machida, H. Kado, K. Yokoyama, and T. Tohda, "Molecular-resolution images of aspirin crystals with atomic force microscopy," Ultramicroscopy **42-44**, 1148 (1992).

C. M. Mate, G. M. McClelland, R. Erlandsson, and S. Chiang, "Atomic-scale friction of a tungsten tip on a graphite surface," Phys. Rev. Lett. **59**, 1942 (1987).

C. M. Mate, R. Erlandsson, G. M. McClelland, and S. Chiang, "Atomic force microscopy studies of frictional forces and of force effects in scanning tunneling microscopy," J. Vac. Sci. Technol. A **6**, 575 (1988).

C. M. Mate, R. Erlandsson, G. M. McClelland, and S. Chiang, "Direct measurement of forces during scanning tunneling microscope imaging of graphite," Surf. Sci. **208**, 208 (1989a).

C. M. Mate, M. R. Lorenz, and V. J. Novotny, "Atomic force microscopy of polymeric liquid films," J. Chem. Phys. **90**, 7550 (1989b).

C. M. Mate, M. R. Lorenz, and V. J. Novotny, "Determination of lubricant film thickness on a particulate disk surface by atomic force microscopy," IEEE Trans. Magn. **26**, 1225 (1990).

C. M. Mate, "Atomic-force-microscope study of polymer lubricants on silicon surfaces," Phys. Rev. Lett. **68**, 3323 (1992).

J. R. Matey and J. Blanc, "Scanning capacitance microscopy," J. Appl. Phys. **57**, 1437 (1985).

G. M. McClelland, R. Erlandsson, and S. Chiang, "Atomic force microscopy: general principles and a new implementation," in *Review of Progress in Quantitative Nondestructive Evaluation*, edited by D. O. Thompson and D. E. Chimenti (Plenum, New York, 1987) p. 307.

G. M. McClelland and S. R. Cohen, "Tribology at the atomic scale," Chem. and Phys. Solid Surf. VIII, edited by R. Vanselow and R. Rowe (Springer, Berlin, 1990).

S. C. Meepagala, F. Real, and C. B. Reyes, "Measurement of tip-sample interaction forces in scanning tunneling microscopy," Proceedings of the STM'90/NANO I Conference, Baltimore, MD, July 23-27, 1990.

E. Meyer, H. Heinzelmann, P. Grütter, Th. Jung, Th. Weisskopf, H.-R. Hidber, R. Lapka, H. Rudin, and H.-J. Güntherodt, "Comparative study of lithium fluoride and graphite by atomic force microscopy (AFM)," J. Micros. **151**, 269 (1988).

E. Meyer, H. Heinzelmann, P. Grütter, Th. Jung, H.-R. Hidber, H. Rudin, and H.-J. Güntherodt, "Atomic force microscopy for the study of tribology and adhesion," Thin Solid Films **181**, 527 (1989a).

E. Meyer, H. Heinzelmann, P. Grütter, Th. Jung, L. Scandella, H.-R. Hidber, H. Rudin, H.-J. Güntherodt, and S. Schmidt, "Investigation of hydrogenated amorphous carbon coatings for magnetic data storage media by atomic force microscopy," Appl. Phys. Lett. **55**, 1624 (1989b).

E. Meyer, H. Heinzelmann, H. Rudin, and H.-J. Güntherodt, "Atomic resolution on LiF(001) by atomic force microscopy," Z. Phys. B **79**, 3 (1990a).

E. Meyer, R. Wiesendanger, D. Anselmetti, H.-R. Hidber, H.-J. Güntherodt, F. Lévy, and H. Berger, "Different response of atomic force microscopy and scanning tunneling microscopy to charge density waves," J. Vac. Sci. Technol. A **8**, 495 (1990b).

E. Meyer, H. Heinzelmann, D. Brodbeck, G. Overney, R. Overney, L. Howald, H.-R. Hidber, and H.-J. Güntherodt, "Atomic resolution on the surface of LiF(100) by atomic force microscopy," Proceedings of the STM'90/NANO I Conference, Baltimore, MD, July 23-27, 1990c.

E. Meyer, L. Howald, R. M. Overney, D. Brodbeck, R. Lüthi, H. Haefke, J. Frommer, and H.-J. Güntherodt, "Structure and dynamics of solid surfaces observed by atomic force microscopy," Ultramicroscopy **42-44**, 274 (1992).

G. Meyer and N. M. Amer, "Erratum: novel optical approach to atomic force microscopy," Appl. Phys. Lett. **53**, 2400 (1988a).

G. Meyer and N. M. Amer, "Novel optical approach to atomic force microscopy," Appl. Phys. Lett. **53**, 1045 (1988b).

G. Meyer and N. M. Amer, "Simultaneous measurement of lateral and normal forces with an optical-beam-deflection atomic force microscope," Appl. Phys. Lett. **57**, 2089 (1990a).

G. Meyer and N. M. Amer, "Optical-beam-deflection atomic force microscopy: the NaCl (001) surface," Appl. Phys. Lett. **56**, 2100 (1990b).

R. O. Miles, A. Dandridge, A. B. Tveten, H. F. Taylor, and T. G. Giallorenzi, "Feedback-induced line broadening in cw channel-substrate planar laser diodes," Appl. Phys. Lett. **37**, 990 (1980).

R. O. Miles, A. Dandridge, A. B. Tveten, and T. G. Giallorenzi, "An external cavity diode laser sensor," J. Lightwave Technol. **LT-1**, 81 (1983).

R. G. Miller and P. J. Bryant, "Atomic force microscopy of layered compounds," J. Vac. Sci. Technol. A 7 (4), 2879 (1989)

T. Miyamoto, R. Kaneko, and S. Miyake, "Tribological characteristics of amorphous carbon films investigated by point contact microscopy," Proceedings of the STM'90/NANO I Conference, Baltimore, MD, July 23-27, 1990.

Y. N. Moiseev, V. M. Mostepanendo, V. I. Panov, and I. Y. Sokolov, "Experimental and theoretical study of the forces and spatial resolution in an atomic-force microscope," Sov. Phys. Tech. Phys. **35**, 84 (1990).

Y. N. Moiseev, V. I. Panov, S. V. Savinov, I. V. Yaminsky, P. Todua, and D. Znamensky, "Atomic force and scanning tunneling microscopy of comb-like cholesteric liquid crystalline polymer LB films," Ultramicroscopy **42-44**, 304 (1992).

R. Möller, A. Esslinger, and B. Koslowski, "Thermal noise in vacuum scanning tunneling microscopy at zero bias voltage," J. Vac. Sci. Technol. A **8**, 590 (1990).

J. Moreland and P. Rice, "High-resolution, tunneling-stabilized magnetic imaging and recording," Appl. Phys. Lett. **57**, 310 (1990a).

J. Moreland and P. Rice, "Tunneling-stabilized magnetic imaging and recording using a scanning tunneling microscope," Proceedings of the STM'90/NANO I Conference, Baltimore, MD, July 23-27, 1990b.

J. Moreland and P. Rice, "Tunneling stabilized magnetic force microscopy: prospects for low temperature application to superconductors," IEEE Trans. Magn. **27**, No. 2, March 1991.

S. Morita, T. Ishizaka, Y. Sugawara, T. Okada, S. Mishima, S. Imai, and N. Mikoshiba, "Surface conductance of metal surfaces in air studied with a force microscope," Jpn. J. Appl. Phys. **28**, L1634 (1989).

P. J. Mulhern, T. Hubbard, C. S. Arnold, B. L. Blackford, and M. H. Jericho, "A scanning force microscope with a fiber-optic-interferometer displacement sensor," Rev. Sci. Instrum. **62**, 1280 (1991).

P. J. Mulhern, B. L. Blackford, M. H. Jericho, G. Southam, and T. J. Beveridge, "AFM and STM studies of the interaction of antibodies with the S-layer sheath of the archaeobacterium *Methanospirillum hungatei*," Ultramicroscopy **42-44**, 1214 (1992).

P. Muralt and D. W. Pohl, "Scanning tunneling potentiometry," Appl. Phys. Lett. **48**, 514 (1986).

P. Muralt, H. Meier, D. W. Pohl, and H. W. M. Salemink, "Scanning tunneling microscopy and potentiometry on a semiconductor heterojunction," Appl. Phys. Lett. **50**, 1352 (1987).

N

G. Neubauer, S. R. Cohen, G. M. McClelland, D. E. Horn, and C. M. Mate, "Force microscopy with bidirectional capacitance sensor," Rev. Sci. Instrum. **61** (7), 1844 (1990).

M. Nonnenmacher, J. Greschner, O. Wolter, and R. Kassing, "Scanning force microscopy with micromachined silicon sensors," Proceedings of the STM'90/NANO I Conference, Baltimore, MD, July 23-27, 1990.

M. Nonnenmacher and H. K. Wickramasinghe, "Optical absorption spectroscopy by scanning force microscopy," Ultramicroscopy **42-44**, 351 (1992).

M. Nonnenmacher, M. O'Boyle, and H. K. Wickramasinghe, "Surface investigations with a Kelvin probe force microscope," Ultramicroscopy **42-44**, 268 (1992).

O

P. I. Oden, L. A. Nagahara, J. J. Graham, J. Pan, N. J. Tao, Y. Li, T. G. Thundat, J. A. DeRose, and S. M. Lindsay, "Atomic force and scanning tunneling microscopy observations of whisker crystals and surface modification on evaporated gold films," Ultramicroscopy **42-44**, 580 (1992).

F. Ohnesorge, W. M. Heckl, W. Häberle, D. Pum, M. Sara, H. Schindler, K. Schilcher, A. Kiener, D. P. E. Smith, U. B. Sleytr, and G. Binnig, "Scanning force microscopy studies of the S-layers from *Bacillus coagulans* E38-66, *Bacillus sphaericus* CCM2177 and of an antibody binding process," Ultramicroscopy **42-44**, 1236 (1992).

L. Olsson, P. Tengvall, R. Wigren, and R. Erlandsson, "Interaction forces between a tungsten tip and methylated SiO_2 surfaces studied with scanning force microscopy," Ultramicroscopy **42-44**, 73 (1992).

T. Oshio, N. Nakatani, Y. Sakai, N. Suzuki, and T. Kataoka, "Atomic force microscope detection system using an optical fiber heterodyne interferometer free from external disturbances," Ultramicroscopy **42-44**, 310 (1992).

R. M. Overney, L. Howald, J. Frommer, E. Meyer, D. Brodbeck, and H.-J. Güntherodt, "Molecular surface structure of organic crystals observed by atomic force microscopy," Ultramicroscopy **42-44**, 983 (1992).

P

S. M. Paik, S. Kim, and I. K. Schuller, "Molecular dynamics simulation of atomic force microscope," Proceedings of the STM'90/NANO I Conference, Baltimore, MD, July 23-27, 1990.

R. C. Palmer, E. J. Denlinger, and H. Kawamoto, "Capacitive-pickup circuitry VideoDiscs," RCA Rev. **43**, 194 (1982).

Sang-il Park and C. F. Quate, "Theories of the feedback and vibration isolation systems for the scanning tunneling microscope," Rev. Sci. Instrum. **58**, 2004 (1987).

R. Patil, S. Kim, E. Smith, and D. Reneker, "Atomic force microscopy of dendritic crystals of polyethylene," Proceedings of the STM'90/NANO I Conference, Baltimore, MD, July 23-27, 1990.

K. Petermann, *Laser Diode Modulation and Noise* (Kluwer Academic Publishers, KTK Scientific Publishers/Tokyo, 1988).

J. B. Pethica, "Interatomic forces in scanning tunneling microscopy: giant corrugations of the graphite surface," Phys. Rev. Lett. **57**, 3235 (1986).

J. B. Pethica and A. P. Sutton, "On the stability of a tip and flat at very small separations," J. Vac. Sci. Technol. A **6**, 2490 (1988).

J. B. Pethica and A. P. Sutton, "Inelastic flow processes in tip-surface interactions," Proceedings of the STM'90/NANO I Conference, Baltimore, MD, July 23-27, 1990.

Piezo Electric Products, Inc., 186 Massachusetts Ave, Cambridge, MA 02139.

D. W. Pohl, "Some design criteria in scanning tunneling microscopy," IBM J. Res. Dev. **30**, 417 (1986).

C. B. Prater, B. Drake, S. A. C. Gould, H. G. Hansma, and P. K. Hansma, "Scanning ion-conductance microscope and atomic force microscope," (1990).

O. Probst, S. Grafström, J. Kowalski, R. Neumann, and M. Wörtge, "A thermally compensated atomic force microscope for UHV operation," Proceedings of the STM'90/NANO I Conference, Baltimore, MD, July 23-27, 1990.

C. A. J. Putman, B. G. de Grooth, N. F. van Hulst, and J. Greve, "A theoretical comparison between interferometric and optical beam deflection technique for the measurement of cantilever displacement in AFM," Ultramicroscopy **42-44**, 1509 (1992).

R

G. Reis, J. Vancea, H. Wittman, Z. Zweck, and H. Hoffman, "Scanning tunneling microscopy on rough surfaces: tip-shape-limited resolution," J. App. Phys. **67**, 1156 (1990).

M. W. Ribarsky and Uzi Landman, in *Approaches to Modeling of Friction and Wear*, edited by F. F. Ling and C. H. T. Pan (Springer-Verlag, New York, 1988).

W. P. Robins, *Phase Noise in Signal Sources (Theory and Applications)* (Peter Peregrinus Ltd., London, 1982).

K.-H. Robrock, K. N. Tu, D. W. Abraham, and J. B. Clabes, "Study of planarization of cobalt silicide lines and silicon surfaces by scanning force microscopy and scanning electron microscopy," Appl. Phys. Lett. **54**, 1543 (1989).

L. Ruan, C. Bai, Z. Hu, and M. Huang, "Imaging nitronyl nitroxide with atomic force microscope," Proceedings of the STM'90/NANO I Conference, Baltimore, MD, July 23-27, 1990.

D. Rugar, H. J. Mamin, R. Erlandsson, J. E. Stern, and B. D. Terris, "Force microscope using fiber-optic displacement sensor," Surf. Sci. **59**, 2337, (1988).

D. Rugar, H. J. Mamin, and P. Guethner, "Improved fiber-optic interferometer for atomic force microscopy," Appl. Phys. Lett. **55**, 2588 (1989).

D. Rugar, H. J. Mamin, P. Guethner, S. E. Lambert, J. E. Stern, I. McFadyen, and T. Yogi, "Magnetic force microscopy: general principles and application to longitudinal recording media," submitted to J. Appl. Phys. **68**, 1169 (1990).

D. Rugar and P. Hansma, "Atomic force microscopy," preprint (1990).

D. Rugar and P. Grütter, "Squeezing the thermal vibrations of a microcantilever," Proceedings of the STM'90/NANO I Conference, Baltimore, MD, July 23-27, 1990.

D. Rugar and P. Grütter, "Mechanical parametric amplification and thermomechanical noise squeezing," Phys. Rev. Lett. **67**, 699 (1991).

D. Rugar, C. S. Yanoni, and J. A. Sidles, "Mechanical detection of magnetic resonance," Nature **360**, 563 (1992).

S

J. J. Sáenz, N. García, P. Grütter, E. Meyer, H. Heinzelmann, R. Wiesendanger, L. Rosenthaler, H.-R. Hidber, and H.-J. Güntherodt, "Observation of magnetic forces by the atomic force microscope," J. Appl. Phys. **62**, 4293 (1987).

J. J. Sáenz, N. García, and J. C. Slonczewski, "Theory of magnetic imaging by force microscopy," Appl. Phys. Lett. **53**, 1449 (1988).

D. Sarid, D. Iams, V. Weissenberger, and L. S. Bell, "Compact scanning-force microscope using a laser diode," Opt. Lett. **13**, 1057 (1988).

D. Sarid, V. Weissenberger, D. Iams, and J. T. Ingle, "Theory of the laser diode interaction in scanning force microscopy," IEEE J. Quantum Electron. **25**, 1968 (1989a).

D. Sarid, D. Iams, J. Ingle, V. Weissenberger, and Josef Ploetz, "Performance of a scanning force microscope using a laser diode," J. Vac. Sci. Technol. A **8**, 378 (1989b).

D. Sarid, D. Iams, J. T. Ingle, V. Weissenberger, and J. Ploetz, "Performance of a scanning force microscope using a laser diode," J. Vac. Sci. Technol. A **8**, 378 (1990).

D. Sarid, "Review of scanning force microscopy," Proceedings of the STM'90/NANO I Conference, Baltimore, MD, July 23-27, 1990.

D. Sarid and V. Elings, "Review of scanning force microscopy," J. Vac. Sci. Technol. B **9**, 431 (1991).

D. Sarid, P. Pax, L. Yi, S. Howells, M. Gallagher, T. Chen, V. Elings, and D. Bocek, "Improved atomic force microscope using a laser diode interferometer," Rev. Sci. Instrum. **63**, 3905 (1992a).

D. Sarid, T. Chen, S. Howells, M. Gallagher, L. Yi, D. Lichtenberger, Nebesney, Ray, D. Huffman, and L. Lamb, "Buckyball-substrate interactions probed by STM and AFM," Ultramicroscopy **42-44**, 610 (1992b).

F. Saurenbach and B. D. Terris, "Imaging of ferroelectric domain walls by force microscopy," Appl. Phys. Lett. **56**, 1703 (1990).

F. Schabert, A. Hefti, K. Goldie, A. Stemmer, A. Engel, E. Meyer, R. M. Overney, and H.-J. Güntherodt, "Ambient-pressure scanning probe microscopy of 2D regular protein arrays," Ultramicroscopy **42-44**, 1118 (1992).

H. Schmidt, J. Heil, J. Wesner, and W. Grill, "Atomic force sensors constructed from carbon and quartz fibers," J. Vac. Sci. Technol. A **8**, 388 (1990).

J. Schneir, O. Marti, G. Remmers, D. Gläser, R. Sonnenfeld, B. Drake, P. K. Hansma, and V. Elings, "Scanning tunneling microscopy and atomic force microscopy of the liquid-solid interface," J. Vac. Sci. Technol. A **6**, 283 (1988).

C. Schönenberger and S. F. Alvarado, "A differential interferometer for force microscopy," Rev. Sci. Instrum. **60**, 3131 (1989).

C. Schönenberger and S. F. Alvarado, "Understanding magnetic force microscopy," Z. Phys. B **80**, 373 (1990a).

C. Schönenberger and S. F. Alvarado, "Observation of single charge carriers by force microscopy," Phys. Rev. Lett. **65**, 3162 (1990b).

C. Schönenberger, S. F. Alvarado, S. E. Lambert, and I. L. Sanders, "Separation of magnetic and topographic effects in force microscopy," J. Appl. Phys. **67**, 7278 (1990).

J. E. Shigley, *Mechanical Engineering Design* (McGraw-Hill Book Company, New York, 1963).

J. A. Sidles, J. L. Garbini, and G. P. Drobny, "The theory of oscillator-coupled magnetic resonance with potential applications to molecular imaging," Rev. Sci. Instrum. **63**, 3881 (1992).

E. J. Snyder, M. S. Anderson, W. M. Tong, R. S. Williams, S. J. Anz, M. M. Alvarez, Y. Rubin, F. N. Diederich, and R. L. Whetten, "Atomic force microscope studies of fullerene films: highly stable C_{60} fcc (311) free surfaces," Science **253**, 171 (1991).

J. M. Soler, A. M. Baró, N. García, and H. Rohrer, "Interatomic forces in scanning tunneling microscopy: giant corrugations of the graphite surface," Phys. Rev. Lett. **57**, 444 (1986).

M. Stedman, "Limits of topographic measurement by the scanning tunneling and atomic force microscopes," J. Micros. **152**, 611 (1988).

J. E. Stern, B. D. Terris, H. J. Mamin, and D. Rugar, "Deposition and imaging of localized charge on insulator surfaces using a force microscope," Appl. Phys. Lett. **53**, 2717 (1988).

F. H. Stillinger and T. A. Weber, "Computer simulation of local order in condensed phases of silicon," Phys. Rev B **31**, 5262 (1985).

W. Stocker, B. Bickmann, S. N. Magonov, H.-J. Cantow, B. Lotz, J.-C. Wittmann, and M. Möller, "Surface structure on polymers and their model compounds observed by atomic force microscopy," Ultramicroscopy **42-44**, 1141 (1992).

K. Sueoka, K. Okuda, N. Matsubara, and F. Sai, "Study of tip magnetization behavior in MFM," Proceedings of the STM'90/NANO I Conference, Baltimore, MD, July 23-27, 1990.

Y. Sugawara, T. Ishizaka, S. Morita, S. Imai, and N. Mikoshiba, "Simultaneous observation of atomically resolved AFM/STM images of a graphite surface," Jpn. J. Appl. Phys. **29**, L296 (1990a).

Y. Sugawara, T. Ishizaka, and S. Morita, "Scanning force/tunneling microscopy (AFM/STM) of a graphite surface in air," Proceedings of the STM'90/NANO I Conference, Baltimore, MD, July 23-27, 1990b.

Y. Sugawara, Y. Fukano, Y. Kamihara, S. Morita, A. Nakano, T. Ida, and R. Kaneko, "AFM/STM investigation of polycrystalline Si surface," Ultramicroscopy **42-44**, 1372 (1992).

A. P. Sutton and J. B. Pethica, "Inelastic flow processes in nanometre volumes of solids," J. Phys. **2**, 5317 (1990).

M. Suzuki, Y. Kudoh, Y. Homma, and R. Kaneko, "Monoatomic step observation on Si(111) surfaces by force microscopy in air," Appl. Phys. Lett. **58**, 2225 (1991).

T

M. Tabib-Azar, "Optically controlled silicon microactuators," Nanotechnology **1**, 81 (1990).

S. L. Tang, J. Bokor, and R. H. Storz, "Direct force measurement in scanning tunneling microscopy," Appl. Phys. Lett. **52**, 188 (1988).

J. Tansock and C. C. Williams, "Force measurement with a piezoelectric cantilever in a scanning force microscope," Ultramicroscopy **42-44**, 1464 (1992).

M. A. Taubenblatt, "Lateral forces and topography using scanning tunneling microscopy with optical sensing of the tip position," Appl. Phys. Lett. **54**, 801 (1989).

E. C. Teague, F. E. Scire, S. M. Baker, and S. W. Jensen, "Three-dimensional stylus profilometry," Thin Solid Films (1982).

B. D. Terris, J. E. Stern, D. Rugar, and H. J. Mamin, "Novel study of contact electrification using force microscopy," Solid State Phys. (1989a).

B. D. Terris, J. E. Stern, D. Rugar, and H. J. Mamin, "Contact electrification using force microscopy," Phys. Rev. Lett. **63**, 2669 (1989b).

B. D. Terris, J. E. Stern, D. Rugar, and H. J. Mamin, "Localized charge force microscopy," J. Vac. Sci. Technol. A **8**, 374 (1990a).

B. D. Terris, F. Saurenbach, R. Twieg, and Nguyen, "Force microscopy of liquid crystal surfaces," Proceedings of the STM'90/NANO I Conference, Baltimore, MD, July 23-27, 1990b.

T. Thundat, D. P. Allison, R. J. Warmack, and T. L. Ferrell, "Imaging isolated strands of DNA molecules by atomic force microscopy," Ultramicroscopy **42-44**, 1101 (1992).

S. Timoshenko and J. Goodier, *Theory of Elasticity*, 3rd ed. (McGraw-Hill Book Company, New York, 1970).

M. Tortonese, R. C. Barrett, and C. F. Quate, "Atomic resolution with an atomic force microscope using piezoresistive detection," Appl. Phys. Lett. **62**, 835 (1993).

U

N. Umeda, H. Uwai, and S. Ishizaki, "Scanning attractive force microscope using photothermal vibration," Proceedings of the STM'90/NANO I Conference, Baltimore, MD, July 23-27, 1990.

V

J. Vesenka, M. Guthold, C. L. Tang, D. J. Keller, E. Delaine, and C. Bustamante, "Substrate preparation for reliable imaging of DNA molecules with the scanning force microscope," Ultramicroscopy **42-44**, 1243 (1992).

A. P. Volodin and M. V. Marchevsky, "Magnetic force microscopy investigation of superconductors: first results," Ultramicroscopy **42-44**, 757 (1992).

W

A. Wadas, "The theoretical aspect of atomic force microscopy used for magnetic materials," J. Magn. Mater. **71**, 147 (1988a).

A. Wadas, "Magnetic forces measured by atomic force microscopy: theoretical approach," J. Magn. Mater. **72**, 295 (1988b).

A. Wadas, "Description of magnetic imaging in atomic force microscopy," J. Magn. Magn. Mat. **78**, 263 (1989).

A. Wadas and P. Grütter, "Theoretical approach to magnetic force microscopy," Phys. Rev. B **39**, 12013 (1989).

A. Wadas and H.-J. Güntherodt, "Lateral resolution in magnetic force microscopy: applications to periodic structures," Phys. Lett. **146**, 277 (1990).

A. Wadas, P. Grütter, and H.-J. Güntherodt, "Analysis of in-plane bit structure by magnetic force microscopy," J. Appl. Phys. **67**, 3462 (1990a).

A. Wadas, P. Grütter, and H.-J. Güntherodt, "Analysis of magnetic bit pattern by magnetic force microscopy," J. Vac. Sci. Technol. A **8**, 416 (1990b).

A. Wadas, H. Hug, E. Meyer, and H.-J. Güntherodt, "The influence from topography on magnetic force microscopy signal," Proceedings of the STM'90/NANO I Conference, Baltimore, MD, July 23-27, 1990c.

S. Watanabe, K. Hane, and T. Goto, "Force microscope using a twin-path interferometer," J. Vac. Sci. Technol. **10**, 1 (1992).

O. Watanuki, F. Sai, and K. Sueoka, "Magnetic-force-sensing STM: novel application of STM for simultaneous measurement of topography and field gradient of magnetic recording heads," Ultramicroscopy **42-44**, 315 (1992).

J. M. R. Weaver and D. W. Abraham, "High resolution AFM potentiometry," Proceedings of the STM'90/NANO I Conference, Baltimore, MD, July 23-27, 1990.

J. M. R. Weaver and H. K. Wickramasinghe, "Semiconductor characterization by SFM surface photovoltage microscopy," Proceedings of the STM'90/NANO I Conference, Baltimore, MD, July 23-27, 1990.

T. P. Weihs, Z. Nawaz, S. P. Jarvis, and J. B. Pethica, "Limits of imaging resolution for atomic force microscopy of molecules," Appl. Phys. Lett. **59**, 3536 (1991).

A. L. Weisenhorn, P. K. Hansma, T. R. Albrecht, and C. F. Quate, "Forces in atomic force microscopy in air and water," Appl. Phys. Lett. **54**, 2651 (1989).

A. L. Weisenhorn, J. E. MacDougall, S. A. C. Gould, S. D. Cox, W. S. Wise, J. Massie, P. Maivald, V. B. Elings, G. D. Stucky, and P. K. Hansma, "Imaging and manipulating molecules on a zeolite surface with an atomic force microscope," Science **247**, 1330 (1990a).

A. L. Weisenhorn, H. G. Hansma, H. E. Gaub, R. L. Sinsheimer, S. A. C. Gould, and P. K. Hansma, "Progress in sequencing DNA with an atomic force microscope," Proceedings of the STM'90/NANO I Conference, Baltimore, MD, July 23-27, 1990b.

A. L. Weisenhorn, M. Egger, F. Ohnesorge, S. A. C. Gould, S. P. Heyn, H. G. Hansma, R. L. Sinsheimer, H. E. Gaub, and P. K. Hansma, "Molecular-resolution images of Langmuir-Blodgett films and DNA by atomic force microscopy," Langmuir **7**, 8 (1991).

A. L. Weisenhorn, F.-J. Schmitt, W. Knoll, and P. K. Hansma, "Streptavidin binding observed with an atomic force microscope," Ultramicroscopy **42-44**, 1125 (1992).

D. J. Whitehouse, "Dynamic aspects of scanning surface instruments and microscopes," Nanotechnology **1**, 93 (1990).

H. Kumar Wickramasinghe, "Scanned probe microscopies," Sci. Am. **261**, 74 (1989).

H. Kumar Wickramasinghe, "Scanning probe microscopy: current status and future trends," J. Vac. Sci. Technol. A **8**, 363 (1990).

R. Wiesendanger, H.-J. Güntherodt, G. Güntherodt, R. J. Gambino, and R. Ruf, "Observation of vacuum tunneling of spin-polarized electrons with the scanning tunneling microscope," Phys. Rev. Lett. **65**, 247 (1990).

R. Wiesendanger, I. V. Shvets, D. Bürgler, G. Tarrach, H.-J. Güntherodt, and J. M. D. Coey, "Recent advances in spin-polarized scanning tunneling microscopy," Ultramicroscopy **42-44**, 338 (1992).

C. C. Williams, W. P. Hough, and S. A. Rishton, "Scanning capacitance microscopy on a 25 nm scale," Appl. Phys. Lett. **55**, 203 (1989).

K. D. Wise and B. G. Orr, "Micromachined silicon sensors: extending instrumentation systems into the nano world," Proceedings of the STM'90/NANO I Conference, Baltimore, MD, July 23-27, 1990.

O. Wolter, Th. Bayer, and J. Greschner, "Micromachined silicon sensors for scanning force microscopy," Proceedings of the STM'90/NANO I Conference, Baltimore, MD, July 23-27, 1990.

D. L. Worcester, R. G. Miller, and P. J. Bryant, "Atomic force microscopy of purple membranes," J. Micros. **152**, 817 (1988).

D. L. Worcester, H. S. Kim, R. G. Miller, and P. J. Bryant, "Imaging bacteriorhodopsin lattices in purple membranes with atomic force microscopy," J. Vac. Sci. Technol. A **8**, 403 (1990).

X

H. Ximen and P. E. Russell, "Microfabrication of AFM tips using focused ion and electron beam techniques," Ultramicroscopy **42-44**, 1526 (1992).

Y

H. Yamada, T. Fijii, and K. Nakayama, "Experimental study of forces between a tunneling tip and graphite," J. Vac. Sci. Technol. A **6**, 293 (1988).

H. Yamada, S. Akamine, and C. F. Quate, "Imaging of organic molecular films with the atomic force microscope," Ultramicroscopy **42-44**, 1044 (1992).

R. Yang, R. Miller, and P. J. Bryant, "Atomic force profiling by utilizing contact forces," J. Appl. Phys. **63**, 570 (1988).

L. Yi, M. Gallagher, S. Howells, T. Chen, and D. Sarid, "Combined STM-AFM for magnetic applications," AIP Conference Proceedings on Scanned Probe Microscopies **241**, 537 (1992).

R. Young, J. Ward, and F. Scire, "The topografiner: an instrument for measuring surface microtopography," Rev. Sci. Instrum. **43**, 999 (1972).

Z

G. W. Zajac, M. Q. Patterson, P. M. Burrell, and C. Metaxas, "Scanning probe microscopy studies of isotactic polypropylene," Ultramicroscopy **42-44**, 998 (1992).

J. A. N. Zasadzinski and P. K. Hansma, "Scanning tunneling microscopy and atomic force microscopy of biological surfaces," Biochemical Engineering Conference (1988).

J. A. N. Zasadzinski, Jason Schneir, John Gurley, Virgil Elings, and Paul K. Hansma, "Scanning tunneling microscopy of freeze-fracture replicas of biomembranes," Science **239**, 1013 (1988).

Zenhausern, M. Adrian, R. Emch, M. Taborelli, M. Jobin, and P. Descouts, "Scanning force microscopy and cryo-electron microscopy of tobacco mosaic virus as a test specimen," Ultramicroscopy **42-44**, 1168 (1992).

W. Zhong and D. Tománek, "First-principles theory of atomic-scale friction," Phys. Rev. Lett. **64**, 3054 (1990).

W. Zhong, G. Overney, and D. Tománek, "Theory of elastic tip-surface interactions in atomic force microscopy," Proceedings of the STM'90/NANO I Conference, Baltimore, MD, July 23-27, 1990.

W. Zimmermann-Edling, H.-G. Busmann, H. Sprang, and I. V. Hertel, "Imaging polycrystalline CVD diamond films on the micrometer and nanometer scale by STM and AFM," Ultramicroscopy **42-44**, 1366 (1992).

Index